Student Companion and Problems Book for
Biochemistry
CAMPBELL

William M. Scovell
Bowling Green State University

SAUNDERS COLLEGE PUBLISHING

SAUNDERS COLLEGE PUBLISHING
Harcourt Brace College Publishers

Fort Worth Philadelphia San Diego New York Orlando
Austin San Antonio Toronto Montreal London Sydney Tokyo

Copyright © 1995 by Harcourt Brace & Company
Copyright © 1991 by Saunders College Publishing

All rights reserved. No part of this publication may be reproduced or transmitted in any form or by any means, electronic or mechanical, including photocopy, recording, or any information storage and retrieval system, without permission in writing from the publisher.

Requests for permission to make copies of any part of the work should be mailed to: Permissions Department, Harcourt Brace & Company, 6277 Sea Harbor Drive, Orlando, Florida 32887-6777.

Printed in the United States of America.

Scovell: Student Companion and Problems Book for Biochemistry, Second Edition by Campbell

ISBN 0-03-001874-9

7 8 9 0 1 2 3 4 5   021   11 10 9 8 7 6 5 4 3

PREFACE

I hear, and I forget
I see, and I remember
I do, and I understand

"Ancient Chinese Proverb"

The study and understanding of biochemistry are daunting, yet stimulating experiences in the chemistry-biology curriculum. Biochemistry draws on your comprehension of many aspects of fundamental chemistry, introduces new concepts and techniques, and then integrates them into a basic understanding of the activities of a living cell. The intellectual challenge is apparent. As with any new undertaking you must decide on a strategy to accomplish the objectives before you. At the same time, it is beneficial to have some "marker" to evaluate your progress. It is with this in mind that The Student Companion and Problems Book was written. This book accompanies and complements the text, with the objective of helping the student gain a more comprehensive understanding of biochemistry.

Each chapter starts with an overview that highlights the ideas and concepts presented in the corresponding chapter in Biochemistry, second edition by Mary K. Campbell. The *Chapter Objectives* focus attention on specific concepts, techniques, and procedures, as well as approaches to problem solving.

A major emphasis of the book is the collection of *Exercises*, which are accompanied by detailed explanations and answers. It is important that you try to solve these problems before consulting the answers. A major effort and commitment at this point in the learning process pays huge dividends in reinforcing and crystallizing your knowledge.

A *Review of Important Organic Reactions in Biochemistry* is included at the end of Chapter 2. This section emphasizes that biochemistry is clearly "the chemistry of living systems" and that certain chemical reactions provide the foundation for the anabolic and catabolic pathways in cellular metabolism.

In addition, a number of important topics that are currently experiencing rapid, if not explosive, development are "SPOTLIGHTED" in this <u>Student Companion</u>. These essays are found in appropriate chapters and serve to point out the impact that these developments have on today's burgeoning understanding of the molecular basis of biology.

I welcome comments from students and professors, especially those which bring any errors in the text to my attention. Best of luck in your study of biochemistry.

Comments should be addressed to:

Professor William M. Scovell
Department of Chemistry
Overman Hall
Bowling Green State University
Bowling Green, OH 43403

TABLE OF CONTENTS

Chapter	Title	Page
Part I	**An Introduction To Biochemistry**	
1	Biochemistry and the Organization of Cells	1
2	Water: The Solvent for Biochemical Reactions	8
	A REVIEW OF IMPORTANT ORGANIC REACTIONS IN BIOCHEMISTRY	19
Part II	**Components of Cells: Structure and Function**	
3	Amino Acids and Peptides	32
4	The Three-Dimensional Structure of Proteins	45
5	The Behavior of Proteins: Enzymes	60
6	Nucleic Acids: How Structure Conveys Information	72
7	Nucleic Acid Biotechnology Techniques **SPOTLIGHT:** DNA in Court: "Molecular Fingerprinting" Evidence	89
8	Lipids and Membranes **SPOTLIGHT:** Controversial Drugs: Synthetic Steroids	104
Part III	**Energetics and Metabolism: Carbohydrates, Lipids, and Compounds of Nitrogen**	
9	The Importance of Energy Changes and Electron Transfer	127

10	Carbohydrates	144
11	Glycolysis	156
12	The Citric Acid Cycle	166
13	Electron Transport and Oxidative Phosphorylation	178
14	Further Aspects of Carbohydrate Metabolism	191
15	Lipid Metabolism **SPOTLIGHT:** Atherosclerosis: Cholesterol and the LDL Connection	200
16	Photosynthesis	213
17	The Metabolism of Nitrogen	225
18	Metabolism in Perspective	240

Part IV   Workings of the Genetic Code

19	Biosynthesis of Nucleic Acids: Replication and Transcription of the Genetic Code **SPOTLIGHT:** RNA as an Enzyme: The Changing of a Mind-Set	252
20	Protein Synthesis: Translation of the Genetic Message **SPOTLIGHT:** Cystic Fibrosis: The Gene, the Protein and Present Day Candidate For Clinical Gene Therapy Trials	263

1

Biochemistry and the Organization of Cells

This first chapter presents some of the most significant tenets for understanding the chemistry in biological systems - biochemistry. This chapter highlights the unifying principles of biochemistry, which will be developed in more detail in the subsequent chapters. Theories are addressed regarding the origin and character of the earth, the molecules first formed, the possible routes by which life began and the subsequent development of aerobic organisms. Living organisms are made-up of and powered by lifeless, abiotic molecules, which were originally thought to have formed under anaerobic conditions. All living organisms may be classified as either the more primitive single-celled **prokaryote** or the more complex **eukaryote**, which may be single-celled or multicellular. Eukaryotic organisms contain a nucleus and other organelles, while prokaryotic cells lack these subcellular compartments. The different types of organelles, all of which are enveloped by a membrane, provide compartments for specialized metabolic activities, such as the oxidation of ingested foods, replication of DNA, and photosynthesis.

Chapter 1 Biochemistry and the Organization of Cells

The orchestration of these individual metabolic processes and their regulation are expressed as normal cellular activity. The concept of **symbiosis** provides a theoretical basis for the rise of eukaryotic organisms and exemplifies how different organisms may be of mutual benefit to each other.

CHAPTER OBJECTIVES

1. Discuss the theories regarding the origin of life, with specific attention to the molecules that were thought to be the first ones formed.
2. List the unique features of prokaryotic and eukaryotic organisms.
3. List the common features of prokaryotic and eukaryotic organisms.
4. Define what constitutes an organelle and specify the functions of the organelles found in a human cell.
5. Indicate the difference between a cell membrane and a cell wall.
6. Discuss the structural components and the function of mitochondria.
7. Discuss the structural features of the two forms of endoplasmic reticulum and indicate how they differ functionally.
8. Classify some of the organisms mentioned in the chapter in the five-kingdom system.
9. Discuss mutualism and parasitic symbiosis and present an example of each.
10. Discuss the endosymbiotic theory and suggest how it may provide the basis for the origin of eukaryotic cells.

EXERCISES

1. The early atmosphere on earth is thought to be _____ in character, with little or no free _____ .

Chapter 1 Biochemistry and the Organization of Cells

2. Two common features of living organisms are:
   _____
   _____

3. The most abundant element in the universe is _____ , while the two most abundant elements in living organisms are _____ and _____ . In both the universe and in living systems, the non-metallic elements, H, C, N and O, are in _____ (high, low) abundance.

4. Photosynthesis in green plants involves the reaction of water and carbon dioxide to produce carbohydrate (for example, $C_6H_{12}O_6$) and molecular oxygen. Write the balanced equation for this reaction.

5. a. What are the molecular components of a cell membrane and a cell wall?
   b. Are cell membranes and cell walls found in prokaryotic and eukaryotic cells?
   c. What is the function of the cell wall and the membrane?

6. Name several organelles found in plants, but not in animal cells. Indicate their function.

7. How do peroxisomes differ from lysosomes?

8. How does the DNA in bacteria cells differ from the DNA in human cells?

9. List the names of the five major classifications for all organisms. Cite an organism in each classification. State any unique characteristics of each class and the important features common among some or all the classes.

10. Recently it has become possible to take miniscule amounts of DNA and, by a powerful new procedure called the Polymerase Chain Reaction (PCR), make millions of copies of this DNA fragment very efficiently and without laborious cloning procedures. The procedure, however, requires repeated changes in temperature ranging from 20-80°C. The enzyme used in this procedure is the

Chapter 1 Biochemistry and the Organization of Cells

DNA polymerase isolated from the bacterium, *Thermus aquaticus* (Taq), which lives in hot springs at temperatures of 70-80°C.

a. What type of bacterium is the source of this polymerase?
b. Suggest a reason why this particular enzyme is so useful.

11. Indicate which of the following statements are true or false. If a statement is false, correct it.

a. Ribosomes always reside in the nucleus.
b. The mitochondria are organelles in eukaryotic cells which are actively engaged in reductive processes.
c. Chloroplasts, glyoxysomes and mitochondria contain DNA.
d. Methanogens are aerobic bacteria which produce methane from CO_2 and H_2.
e. The mitochondrial proteins are all derived from the genes of mitochondrial DNA.
f. There is clear and convincing evidence that eukaryotic cells arose from a symbiotic relationship of prokaryotic cells.
g. The filaments in the cytoskeleton of prokaryotic cells are primarily made up of the protein tubulin.
h. The Golgi apparatus is involved in the secretion of proteins from the cell.
i. The DNA in a prokaryotic cell is localized primarily in the nucleolus.
j. All proteins have a unique amino acid sequence. Those that exhibit catalytic activity are called enzymes.

ANSWERS TO EXERCISES

1. reducing, oxygen (O_2).

2. All living organisms (1) use energy and (2) make use of the same types of biomolecules.

3. Hydrogen, Carbon, Oxygen; high.

Chapter 1 Biochemistry and the Organization of Cells

4. $$6\ H_2O + 6\ CO_{2(g)} \underset{}{\overset{h\nu}{\rightleftharpoons}} C_6H_{12}O_6 + 6\ O_{2(g)}$$

5. a. Cell membranes are composed of proteins in a double layer (bilayer) of phospholipids. Cell walls consist primarily of polysaccharides.
   b. Both prokaryotic and eukaryotic cells have outer membranes enveloping the cell. Eukaryotic cells also have organelles within the cell which are enclosed by membranes. Prokaryotic cells and plant cells have cell walls, which animal cells do not have.
   c. Cell walls are quite rigid and serve as a protective coating for the cell. The membrane serves as a selective barrier to the intake or release of molecules. In addition, protein molecules that are embedded in the membrane can be involved in intercellular recognition.

6. Green plants have chloroplasts, which contain the apparatus for photosynthesis. Plants also contain glyoxysomes, which contain enzymes used in the glyoxylate cycle. This pathway permits the conversion of lipids to carbohydrates.

7. Peroxisomes are primarily involved with the metabolism (elimination) of toxic hydrogen peroxide, H_2O_2. Lysosomes are sacs of degradative enzymes which are capable of digesting nucleic acids, proteins and lipids.

8. In bacteria, the DNA is a single, circular molecule that contains the genetic information. It is generally localized in the so-called nuclear region, which has no distinct or detectable boundary with the cytoplasm.

   In human cells, the DNA resides in the nucleus, a distinct organelle. The DNA is complexed with nuclear proteins to form chromatin, or individual chromosomes (each chromosome is considered to contain a single molecule of DNA), which are observable under the light or electron microscope. The amount of DNA in a human cell is about 1000 times greater than that found in a bacteria cell.

Chapter 1 Biochemistry and the Organization of Cells

9. The classes are:
   1. Monera — prokaryotic (bacteria)
   2. Protista — unicellular (Paramecium)
   3. Fungi — molds or mushrooms
   4. Plantae — flowers
   5. Animalia — human beings

Class 1 contains only prokaryotic organisms, while the other four classes contain eukaryotic organisms. The Protista class contains primarily unicellular eukaryotic cells, while classes 3-5 are virtually all multicellular. Organisms in the Animalia class do not have cell walls, while those in the Plantae class do have cell walls. All organisms are composed of living cells which must metabolize exogenous nutrients to produce energy and to provide molecules for sustained life.

10.
   a. The polymerase is derived from an archaebacteria called thermacidophile.
   b. The *Thermus aquaticus* bacteria are very hardy in that they survive normally at high temperatures and in some cases, at low pH. This means that their enzymes remain active and stable under these conditions, which would be especially harsh and generally lethal to most other organisms. The DNA polymerase in *Thermus aquaticus* (Taq DNA polymerase) is very stable and active at high temperatures, while DNA polymerases from virtually all other organisms would not function if heated to these extreme temperatures. Therefore, the PCR procedure is carried out with Taq DNA polymerase because it remains active throughout the entire cyclic procedure, especially at the high temperatures that are used.

11.
   a. False. Ribosomes reside in the cytoplasm.
   b. False. Mitochondria are actively engaged in oxidative metabolic processes.
   c. False. Chloroplasts and mitochondria contain DNA. Glyoxysomes do not contain DNA.
   d. False. Methanogens are anaerobic organisms.
   e. False. Some mitochondrial proteins are the products of nuclear genes.
   f. False. The evidence is not compelling.

Chapter 1 Biochemistry and the Organization of Cells

g. False. The filaments in the cytoskeleton occur in eukaryotic cells and they are primarily made up of the protein tubulin.
h. True.
i. False. The DNA in a prokaryotic cell is primarily localized in the nuclear region. The nucleolus is found in the nucleus of a eukaryotic cell.
j. True.

2

Water: The Solvent for Biochemical Reactions

The presence of water on earth and the unique characteristics of water provide the essential basis for life. Its bent structure, polar bonds and capacity to form intermolecular hydrogen bonds makes water exhibit many exceptional properties, including being an unusually high boiling solvent capable of dissolving ions and polar molecules. While compounds that can form hydrogen bonds with water are especially soluble, non-polar or hydrophobic molecules are not soluble. Acids and bases dissolve in water and dissociate either completely (strong acids and bases) or to a small extent (weak acids and bases). The acidity or pH of an aqueous solution is determined by the concentration of protons [H^+] in solution (pH = -log[H^+]). **Buffers**, which are essential to living organisms, are solutions which contain a weak acid and its conjugate base or a weak base and its conjugate acid, in which the pH is equal to or close to the pK value (pK_a' = -logK_a') for the acid or base. The **Henderson-Hasselbalch equation** (pH = pK_a' + log[A^-]/[HA]) readily can be used to calculate

Chapter 2 Water: The Solvent for Biochemical Reactions

the pH of a weak acid or base, a buffered solution or in the calculation of a titration curve.

As a preface for many of the subsequent chapters, the basic functional groups of common organic compounds are presented. It is the functional group (or groups) present in an organic molecule that determines its observed chemical reactivity. As a review, the most common reactions for the functional groups are included in this chapter of the problems book. It should be continuously reviewed and used as a ready source of the general reaction types that are relevant to metabolism and other dynamic aspects of biochemistry.

CHAPTER OBJECTIVES

1. Describe the concept of electronegativity and explain its role in determining whether a bond is polar or non-polar.
2. Explain how ion-dipole and dipole-dipole interactions between dissolved ions or molecules in water aid in their solubility.
3. Define the terms hydrophilic and hydrophobic and discuss how the preferential interactions of molecules with water or with each other determines their solubility or insolubility.
4. Draw the general structure for a micelle and explain the nature of the driving force for micelle formation.
5. Indicate the characteristics a molecule must possess in order to be a hydrogen-bond donor or acceptor.
6. Become familiar with the types of non-covalent bonding forces and the magnitude of the energies for the interactions.
7. State the relationship between the strength of an acid and its K_a' and pK_a' values.
8. Review the relationship between the pH of a solution and its hydrogen-ion concentration.
9. Make a plot for the titration of a weak acid and indicate the species that are present at the beginning, the half-way point and at the equivalence point in the titration.
10. Define the terms buffer and buffer capacity and explain how the buffering capacity changes as a function of pH.
11. Give some examples of buffers and indicate the buffer species that reacts with any added acid or base.

Chapter 2 Water: The Solvent for Biochemical Reactions

12. Review the functional groups in organic molecules, be able to recognize them in complex molecules and know the reactions associated with each functional group.

EXERCISES

1. Predict the relative solubility of the following series of molecules in water. Briefly explain your answer.

    a. CH_3COOH, $CH_3(CH_2)_6COOH$, CH_3CH_2COOH

    b. R-CHO, R-C-R, R-R, R-COOH, R-O-R, in which R = CH_3
    $\parallel$
    O

2. Explain how the solubility of these molecules are affected by pH.

    a. anilinium cation pK_a = 4.70
    b. formic acid pK_a = 3.75
    c. methylacetate   no pK_a value

3. Consider the boiling points for the following molecules, all of which have essentially the same molecular weight. Explain the trend in the boiling points in terms of intermolecular hydrogen bonding.

Molecule	Boiling point (°C)
CH_4 (methane)	-161
NH_3 (ammonia)	-34.5
HF (hydrogen fluoride)	19.4
OH_2 (water)	100

Chapter 2 Water: The Solvent for Biochemical Reactions

4. Consider the molecules listed below and fill in the table. Indicate whether the bond and the molecule is polar or non-polar. Refer to Table 2.1 in the text for electronegativity values.

Molecule	Bonds	Molecule
OH_2	_____	_____
CH_4	_____	_____
CO_2	_____	_____
BF_3	_____	_____

5. a. Indicate whether the molecules listed below are amphiphilic, hydrophilic or hydrophobic.
   b. Define and characterize a micelle.
   c. Indicate which of these molecules can form micelles in an aqueous solution.

Molecule	Character	Micelle Formation(yes/no)
sodium acetate	_____	_____
normal hexane	_____	_____
potassium salt of octanoic acid	_____	_____
naphthalene	_____	_____
SDS (sodium dodecylsulfate)	_____	_____

6. Which of the following molecules interact with each other by hydrogen bonding? For the cases in which this occurs, draw the structure showing the proposed intermolecular interactions.

   a. $CH_3\overset{O}{\overset{\|}{C}}CH_3$ and CH_3OH

   b. Urea, $H_2N\overset{O}{\overset{\|}{C}}NH_2$ and $CH_3CH_2COCH_3$

   c. $CH_3CH_2\overset{O}{\overset{\|}{C}}H$ with itself.

   d. CH_3Cl and NH_3

   e. $N(CH_3)_4^+$ and CH_3COOH

11

Chapter 2 Water: The Solvent for Biochemical Reactions

7. Non-covalent interactions are important in determining the ultimate structure of important biomacromolecules, such as proteins and nucleic acids. Specify the amount of energy (kJ/mol) associated with the following non-covalent interactions.

   Hydrogen bonding   _____
   van der Waals bonds   _____
   Hydrophobic interactions _____

8. Calculate the pH of the following solutions.

   a. 0.1 M pyruvic acid; pK_a = 2.50
   b. 0.1 M lactic acid and 0.1 M sodium lactate; pK_a = 3.86
   c. An aqueous solution containing 2.5×10^{-8} M OH^-.
   d. A 0.1 M succinic acid (pK_{a1} = 4.21; pK_{a2} = 5.63) solution after 0.09 equivalents of NaOH is added.
   e. Which of these solutions (a-d) would serve as an effective buffer?

9. Indicate the conjugate acid and/or base for the following species in aqueous solution.

Molecule	Conjugate base	Conjugate acid
a. H_2O		
b. benzoic acid		
c. NH_3		
d. HCO_3^-		
e. $H_2PO_4^-$		

10. Stomach acid is thought to be primarily concentrated hydrochloric acid (HCl). If 30.0 mL of 0.10 M NaOH is required to neutralize a 50.0 mL sample of stomach acid, calculate the pH of the sample.

Chapter 2 Water: The Solvent for Biochemical Reactions

11. A 0.1 M Tris buffer [tris(hydroxymethyl)aminomethane; pK = 8.1] is adjusted to pH 8.1 and is used as the buffer solution to study an enzymatic reaction which produces 0.01 moles/L of H^+.

    a. What is the initial ratio of [Tris]/[TrisH$^+$]?
    b. Calculate the initial concentration of the TrisH$^+$ and Tris species?
    c. Calculate the concentrations of Tris and TrisH$^+$ after the reaction?
    d. Calculate the final pH of the solution.
    e. Determine the final pH of the solution if the enzymatic reaction were carried out in 1 L of water, pH = 7, with no buffer present.

12. a. Draw a micelle and point out the hydrophobic region and the charged hydrophilic "tails".
    b. Briefly explain the driving force for the formation of a micelle in an aqueous solution.

ANSWERS TO EXERCISES

1. a. $CH_3COOH > CH_3CH_2COOH > CH_3(CH_2)_6COOH$
   These three molecules are simple carboxylic acids; therefore, the relative solubility will decrease as the molecular weight increases. The decreased solubility corresponds to the increase in the molecular weight of the organic (R) group in the acids.
   b. These molecules have approximately the same molecular weight and therefore, the solubility will depend on the extent to which each interacts favorably with water. The extent to which intermolecular hydrogen bonding takes place between the molecule and water is:

   $$R\text{-}R < R\text{-}O\text{-}R < R\text{-}CHO < R\underset{\underset{O}{\|}}{\text{-}C\text{-}}R < R\text{-}COOH$$

   Therefore, the solubility also increases in this order.

Chapter 2 Water: The Solvent for Biochemical Reactions

2. a. The anilinium cation is the protonated form of aniline (pK_a = 4.7). The cation will interact favorably with water and therefore the solubility (of the cation) will increase as the pH progressively is decreased below approximately pH 5. At pH 4.7, there will be equal concentrations of the protonated (soluble) form and the unprotonated aniline (relatively insoluble). As the pH is lowered further, there will be increasing amounts of the protonated, soluble form.

   b. The anion of formic acid, $H-CO_2^-$, will be more soluble than the neutral acid, H-COOH. With a pK_a value of 3.75, its solubility will increase as the pH increases.

   c. This is an ester, which is neither acidic nor basic. Therefore, assuming no reaction, the solubility will be independent of the solution pH. Recall, however, that an ester can be hydrolyzed in strongly basic or acidic solutions.

3. The boiling points for a series of similar molecules are an indication of the extent of **intermolecular** interactions between the molecules. The trend in the boiling points can be explained by the **type** and the **number** of intermolecular interactions between the molecules.
Although all the molecules have weak van der Waals interactions between them, the primary effect on the boiling points and other physical characteristics is the extent of intermolecular hydrogen bonding. CH_4 is not capable of hydrogen bonding. HF can donate one hydrogen and although it has three lone pairs of electrons, HF cannot accept three hydrogens. The strength of the one hydrogen bond in HF is, however, very strong. Each H_2O molecule can donate two hydrogens and the nonbonding-electron pairs in oxygen can accept two hydrogens. This is the optimal hydrogen-bonding capacity, with the number of hydrogen-bond donors equivalent to the number of hydrogen-bond acceptors. NH_3 has three hydrogens which can, in principle, take part in hydrogen bonding, but as with HF, this does not occur. Ammonia is limited to only one intermolecular hydrogen bond. The hydrogen bonds in ammonia are not as strong as in water or

Chapter 2 Water: The Solvent for Biochemical Reactions

hydrogen fluoride because of the smaller difference in the electronegativity value in a (N-H) bond than in a (O-H) or (H-F) bond. One of the key points in explaining the high boiling point for H_2O relative to HF and NH_3, therefore, is that only in water can an extensive hydrogen bonding network occur to optimize intermolecular interactions. This is illustrated in Figure 2.7 in the text.

4.

Molecule	Bonds	Molecule	
OH_2	polar	polar	(bent)
CH_4	non-polar	non-polar	(tetrahedral)
CO_2	polar	non-polar	(linear)
BF_3	polar	non-polar	(trigonal planar)

5. a. and c.

Molecule	Character	Micelle Formation(yes/no)
sodium acetate	hydrophilic	no
normal hexane	hydrophobic	no
potassium salt of octanoic acid	amphiphilic	yes
napthalene	hydrophobic	no
SDS (sodium dodecylsulfate)	amphiphilic	yes

b. A micelle is a globular structure formed by the association of amphiphilic molecules in which the hydrophobic segments reside on the inside of the spherical micelle, while the polar or hydrophilic segments interact with the water on the outer surface of the structure. See Figure 2.3b in the text.

6. a. yes.

$$CH_3-C-CH_3$$
$$\|$$
$$O\ldots\ldots H-O-CH_3$$

b. yes.

Chapter 2 Water: The Solvent for Biochemical Reactions

```
           O
           ‖
     H-N-C-N-H
        |   |
        H   H
        ⋮   ⋮
           O
           ‖
   CH₃CH₂-C-O-CH₃
```

    c.   Aldehydes do not hydrogen bond with each other since there are no hydrogen-bond donors.

    d. NH_3 can act as both a hydrogen-bond donor and acceptor, but CH_3Cl cannot hydrogen bond in any way. There will be no intermolecular hydrogen bonding interactions between these molecules.

    e.   Tetramethylammonium cation does not take part in hydrogen bonding; therefore, there can be no intermolecular hydrogen bonding between this cation and acetic acid, CH_3COOH.

7.

Hydrogen bonding	approximately 20 kJ/mol
van der Waals bonds	approximately 4 kJ/mol
Hydrophobic interactions	4-12 kJ/mol

8.   a.   Pyruvic acid = PyH
            $PyH = Py^- + H^+$ $pK_a = 2.50$
            $K_a = 3 \times 10^{-3} = [Py^-][H^+]/[PyH]$
               $3 \times 10^{-3} = x^2/(0.1 - x)$
               $x = 1.76 \times 10^{-2}$ M $= [Py^-] = [H^+]$.
            pH $= -\log[H^+] = -\log(1.76 \times 10^{-2})$

            <u>pH = 1.76</u>

    b.   Lactic acid = LacH
            $LacH = Lac^- + H^+$ pK = 3.8
            Using the Henderson-Hasselbalch equation,
            pH = pK + log[Lac⁻]/[LacH]
               = 3.86 + log(0.1)/(0.1) = 3.86 + log(1); log 1 = 0

            <u>pH = 3.86</u>

Chapter 2 Water: The Solvent for Biochemical Reactions

c. $[H^+] \times [OH^-] = 1.0 \times 10^{-14}$
If $[OH^-] = 2.5 \times 10^{-8}$ M, $[H^+] = 4.0 \times 10^{-7}$ M

pH = 6.40

d. Use the Henderson-Hasselbalch equation and consider only the first pK_a. The 0.09 equivalents of NaOH converts 90% of the H_2Succ to the $HSucc^-$ form, leaving only 10% of the original H_2Succ.

pH = pK + log $[HSucc^-]/[H_2Succ]$
pH = 4.21 + log (0.9/0.1) = 4.21 + log 9
pH = 4.21 + 0.95

pH = 5.16

e. The only solution which could effectively serve as a buffer is the lactic acid/lactate solution because there are equivalent concentrations of both the acid and its conjugate base.

9.

	Molecule	Conjugate base	Conjugate acid
a.	H_2O	OH^-	H_3O^+
b.	benzoic acid	benzoate anion	none
c.	NH_3	NH_2^-	NH_4^+
d.	HCO_3^-	CO_3^{-2}	H_2CO_3
e.	$H_2PO_4^-$	HPO_4^{-2}	H_3PO_4

10. $(V_a)(M_a) = (V_b)(M_b)$
(30.0 mL) (0.10 M) = (50.0 mL) (x M HCl)
Molarity of HCl = 0.06 M
pH = - log $[H^+]$ = - log $(6.0 \times 10^{-2}$ M) = 2.00 - 0.78

pH = 1.22 for the stomach acid

11. a. The solution pH in this problem is equal to the pK_a value for Tris. Therefore, the ratio of $[Tris]/[TrisH^+]$ is 1.0.

Chapter 2 Water: The Solvent for Biochemical Reactions

b. Since the total buffer concentration is 0.1 M and there is equal concentrations of both the Tris and TrisH$^+$ species, the initial concentration of each species is 0.05 M.

c. The reaction produces 0.01 M H$^+$. The Tris will react with the H$^+$ to increase the concentration of TrisH$^+$ by 0.01 M and to decrease the concentration of Tris by 0.01 M.

The concentrations will be:
[TrisH$^+$] = 0.05 + 0.01 = 0.06 M
[Tris]   = 0.05 - 0.01 = 0.04 M

d. The final pH is calculated with the Henderson-Hasselbalch equation.
pH = 8.1 + log(0.04 M/0.06 M)
pH = 7.92

e. If 0.01 moles/L of hydrogen ion were added to water at pH = 7, the concentration of H$^+$ would be 0.01 M. The pH of a solution with 1.0 x 10^{-2} M H$^+$ is 2. The comparison of the pH values in answers (d) and (e) exemplifies the role of a buffer solution in maintaining the desired pH when carrying out a reaction which produces acid or base.

12. a.

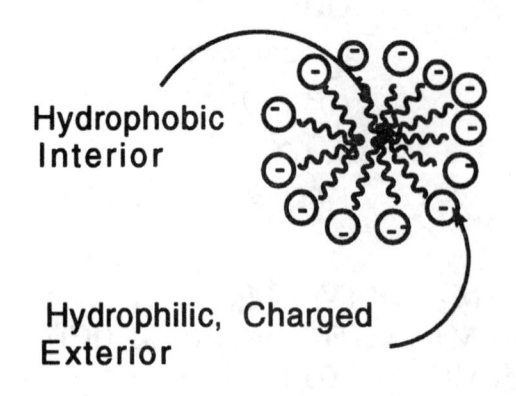

b. The primary driving force for the formation of a micelle is for the hydrophobic segments of the molecules to associate with each other (a favorable interaction) and in the process, not interfere with the structure of water. At the same time, the charged ends point out into the aqueous solution and interact favorably with the water molecules.

Chapter 2 Water: The Solvent for Biochemical Reactions

ADDITIONAL TOPIC:

A REVIEW OF
IMPORTANT ORGANIC REACTIONS IN BIOCHEMISTRY

The "heart and soul" of biochemistry lies in the metabolic processes actively taking place in living cells. These reactions are associated with the enzymatic interconversion of organic molecules. The ability to recognize the organic functional groups in complex molecules and the chemical reactivity associated with each molecule serves as a basis for understanding these processes.

The following summary provides a list of general reaction types that appear in subsequent chapters in the text. The examples are taken directly from organic chemistry. Since these reactions are carried out in non-living systems, they often involve the use of a strong acid or base or the addition of heat to drive the reaction to completion. These same types of reactions are prevalent in biological systems, but are catalyzed by cellular enzymes and occur at physiological conditions. The reactions are cataloged according to the functional group.

1. **Alkenes (Olefins)** (>C = C<)
   Hydration- (addition of water to alkenes to produce an alcohol)

$$-\underset{|}{C}=\underset{|}{C}- \;+\; H_2O \;\xrightarrow{H^+}\; -\underset{|}{\underset{H}{C}}-\underset{|}{\underset{OH}{C}}-$$

2. **Alcohols** R - OH
   a. Dehydration of Alcohols to Yield an Olefin

$$-\underset{|}{\underset{H}{C}}-\underset{|}{\underset{OH}{C}}- \;\xrightarrow{H^+}\; -\underset{|}{C}=\underset{|}{C}- \;+\; H_2O$$

19

Chapter 2 Water: The Solvent for Biochemical Reactions

b. Oxidation of Alcohols (where [O] is a general oxidizing agent)

(i) <u>Primary</u>:

$$RCH_2-OH \xrightarrow{[O]} R-\underset{\underset{O}{\parallel}}{C}-H \xrightarrow{[O]} R-\underset{\underset{O}{\parallel}}{C}-OH$$

aldehyde   carboxylic acid

(ii) <u>Secondary</u>:

$$R-\underset{\underset{OH}{|}}{C}H-R' \xrightarrow{[O]} R-\underset{\underset{O}{\parallel}}{C}-R'$$

ketone

(iii) <u>Tertiary</u>: Tertiary alcohols cannot be oxidized.

3. Thiols   R-SH

a. Oxidation of Thiols to Form a Disulfide Bond.

$$2\ R-S-H \xrightarrow{[O]} R-S-S-R$$

disulfide

b. Reduction of Disulfides (where [R] is a general reducing agent)

$$R-S-S-R \xrightarrow{[R]} 2\ R-S-H$$

4. Aldehydes

$$R-\underset{\underset{O}{\parallel}}{C}-H$$

a. Oxidation

$$R-\underset{\underset{O}{\parallel}}{C}-H \xrightarrow{[O]} R-\underset{\underset{O}{\parallel}}{C}-OH$$

carboxylic acid

Chapter 2 Water: The Solvent for Biochemical Reactions

b. Reduction

$$R\text{-}C(\!=\!O)\text{-}H \xrightarrow{[R]} R\text{-}CH_2\text{-}OH \text{ (primary alcohol)}$$

c. Hemiacetal and Acetal Formation
(reaction of an aldehyde with an alcohol)

```
                                    OR'                  OR'
                                    |      H+/R'OH       |
R-C-H    +   R'-OH   --->    R-C-H   ----------->   R-C-H   + H2O
  ||                           |     <-----------     |
  O                            OH                    OR'
aldehyde    alcohol          hemiacetal             acetal
```

d. Aldol Condensation
(combination of two carbonyl compounds to form an aldol - a ß-hydroxy carbonyl)

```
  O   R3              O                       O   R3  OH
  ||  |               ||                      ||  |   |
R1- C - C - H    +    C- R4    ------>    R1- C - C - C - R4
      |               |        <-----              |   |
      R2              H                            R2  H

ketone             aldehyde                    aldol
                                          (ß-hydroxy carbonyl)
```

e. Reversible Isomerization of Ketose and an Aldose

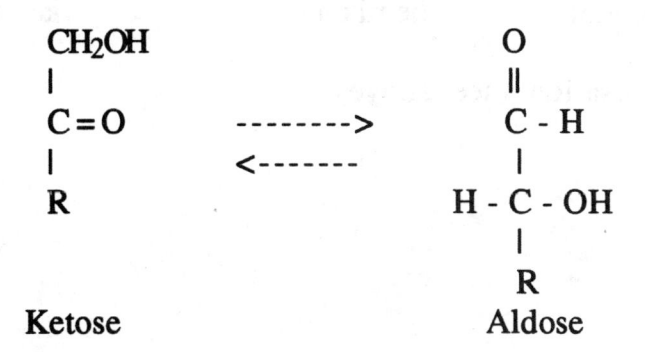

```
  CH2OH                          O
  |                              ||
  C = O        -------->         C - H
  |            <--------         |
  R                              H - C - OH
                                 |
                                 R
Ketose                         Aldose
```

21

Chapter 2 Water: The Solvent for Biochemical Reactions

5. Ketones   R - C - R'
                                           ‖
                                           O

    a. Oxidation - no reaction

    b. Reduction

$$R-\underset{\underset{O}{\|}}{C}-R' \xrightarrow{[R]} R-\underset{\underset{OH}{|}}{\overset{\overset{H}{|}}{C}}-R'$$
$$\text{secondary alcohol}$$

    c. Keto-enol Tautomerism (for guanine)

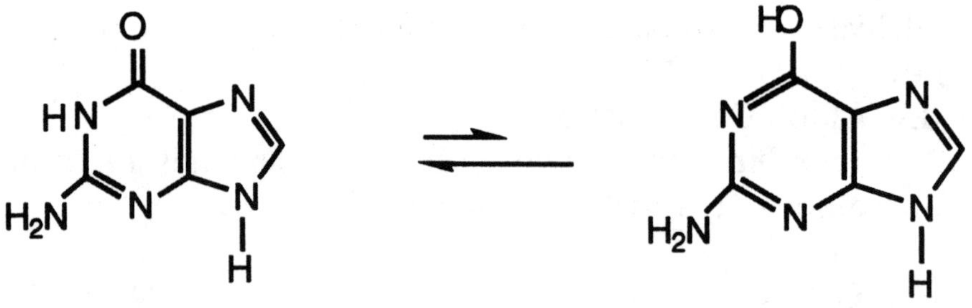

Keto tautomer   Enol tautomer

    d. Hemiketal and Ketal Formation

$$R-\underset{\underset{O}{\|}}{C}-R' + R''-OH \longrightarrow R-\underset{\underset{OH}{|}}{\overset{\overset{OR''}{|}}{C}}-R' \underset{\longleftarrow}{\overset{H^+/R''OH}{\longrightarrow}} R-\underset{\underset{OR''}{|}}{\overset{\overset{OR''}{|}}{C}}-R' + H_2O$$

ketone   alcohol   hemiketal   ketal

    e. Aldol Condensation (see above)

Chapter 2 Water: The Solvent for Biochemical Reactions

6. Carboxylic Acids   R - C - OH
                              ‖
                              O

   a. Reduction

$$R-\underset{\underset{O}{\|}}{C}-OH \xrightarrow{[R]} R-\underset{\underset{O}{\|}}{C}-H \xrightarrow{[R]} R-CH_2-OH$$

                                          aldehyde   primary alcohol

   b. Decarboxylation
      Carboxylic acids with an α-carbonyl group undergo loss of CO_2 (decarboxylation) when heated.

$$HO-\underset{\underset{O}{\|}}{C}-\underset{\underset{O}{\|}}{C}-OH \xrightarrow{\text{Heat}} HO-\underset{\underset{O}{\|}}{C}-H + CO_2$$

7. Esters   R - C - OR'
                    ‖
                    O

   a. Preparation of Esters

$$R-\underset{\underset{O}{\|}}{C}-OH + R'-OH \longrightarrow R-\underset{\underset{O}{\|}}{C}-O-R' + H_2O$$

   b. Hydrolysis (in basic or acidic aqueous solution)

$$R-\underset{\underset{O}{\|}}{C}-O-R' \xrightarrow{OH^-} R-\underset{\underset{O}{\|}}{C}-O^- + R'-OH$$

$$R-\underset{\underset{O}{\|}}{C}-O-R' \xrightarrow{H^+} R-\underset{\underset{O}{\|}}{C}-OH + R'-OH$$

Chapter 2 Water: The Solvent for Biochemical Reactions

    c. Claisen Condensation
      (This ester condensation reaction is similar to the aldol condensation, but here the -OR group of the ester is the leaving group. Therefore, the **Claisen condensation is a substitution reaction**, while the **aldol condensation is an addition reaction**).

```
      O                O                         O   O
      ||               ||          base          ||  ||
 RCH₂C-OR'   +   H-CHCOR'    -------->    RCH₂C-CHCOR'  +  R'OH
                   |                              |
                   R                              R
                                             a β-keto ester
```

8. <u>Amines</u> $NH_nR_{(3-n)}$

Schiff Base Formation

```
        O                              H
        ||              H+             |
   R - C - H  +   R'- NH₂  -------->  R - C = NR'

   carbonyl      substituted          Schiff base
   compound        amine          (substituted imine)
```

9. <u>Amides</u>   R - C = O
                                        |
                                 $NH_nR_{(3-n)}$

    a. Peptide Bond (Amide) Formation

```
  H            H H                    H     H H
  |            | |        DCC         |     | |
H₂N-C-C-O-H + H-N-C-C-OH -----> H₂NC- C - N-CCO₂H + H₂O
  | ||            | ||                |     | ||  |
  R  O            R  O                R     R O   R
```

(DDC = dicyclohexylcarbodiimide, which activates the carboxyl group.)

Chapter 2 Water: The Solvent for Biochemical Reactions

b. Hydrolysis

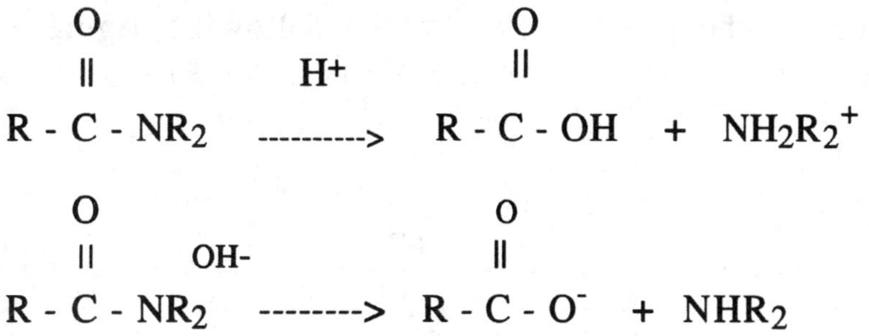

$$R-\underset{\underset{O}{\|}}{C}-NR_2 \xrightarrow{H^+} R-\underset{\underset{O}{\|}}{C}-OH + NH_2R_2^+$$

$$R-\underset{\underset{O}{\|}}{C}-NR_2 \xrightarrow{OH^-} R-\underset{\underset{O}{\|}}{C}-O^- + NHR_2$$

10. **Esters of Phosphoric Acid**

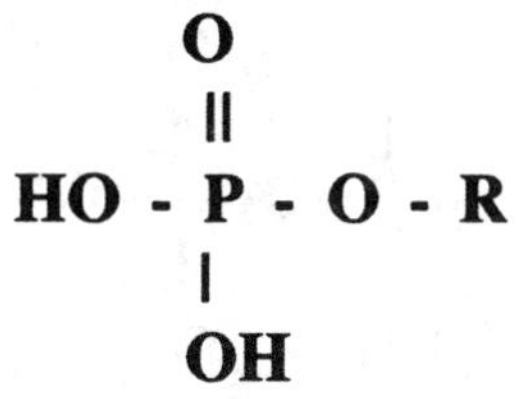

$$HO-\underset{\underset{OH}{|}}{\overset{\overset{O}{\|}}{P}}-O-R$$

Hydrolysis

$$HO-\underset{\underset{OH}{|}}{\overset{\overset{O}{\|}}{P}}-O-R \xrightarrow{H_2O} HO-\underset{\underset{OH}{|}}{\overset{\overset{O}{\|}}{P}}-OH + R-OH$$

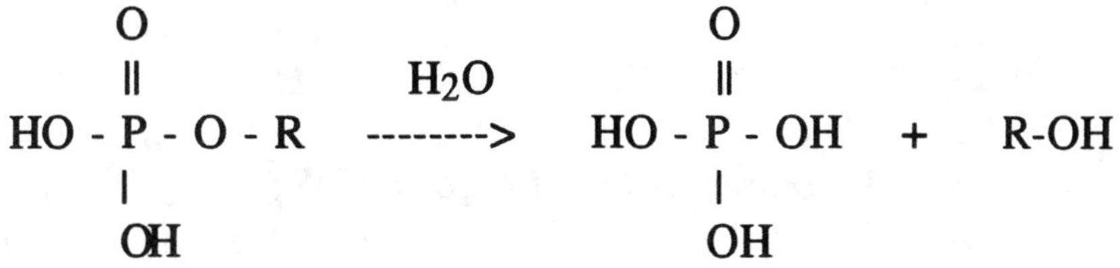

EXERCISES

1. Name all the classes of organic compounds that contain the following:

    a. a carbonyl group.
    b. a nitrogen atom.
    c. a hydroxyl group.

2. Write the equation for the reaction of the following molecules and show the structures for the reactants and products:

    a. butyric acid and propanol

Chapter 2 Water: The Solvent for Biochemical Reactions

    b.   phosphoric acid and ethanol

    c.   phosphoric acid and 5'-adenosine diphosphate (5'-ADP) to produce 5'-adenosine triphosphate (5'-ATP) containing an additional acidic anhydride bond. Hint: The reaction occurs at the point shown in the following figure of 5'-ADP.

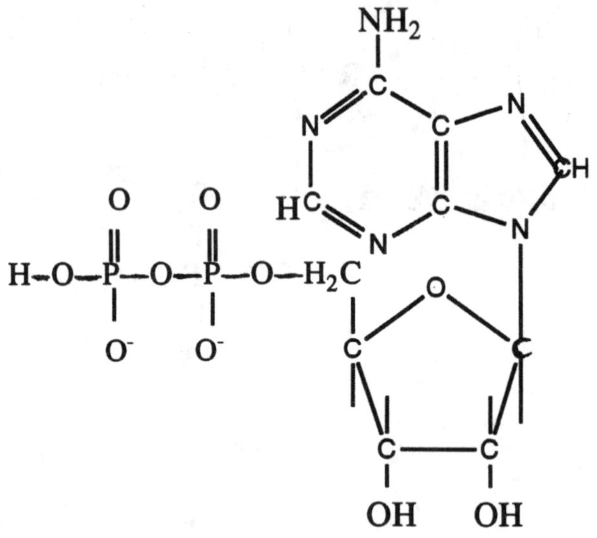

Adenosine diphosphate (ADP)

    d.   Glycerol, $HOCH_2CHOHCH_2OH$, which is a triol, and an excess of the saturated fatty acid, $H_3C(CH_2)_{14}COOH$.

3   a.   The structure for the a-amino acid, lysine, is shown to the right. Indicate the Greek letter which designates the starred (*) carbon connected to the amine group in the side chain?

    b.   Draw the structure for the substituted lysine, 5-hydroxylysine.

Chapter 2 Water: The Solvent for Biochemical Reactions

4. Identify the functional groups in the compounds shown below.

   a.

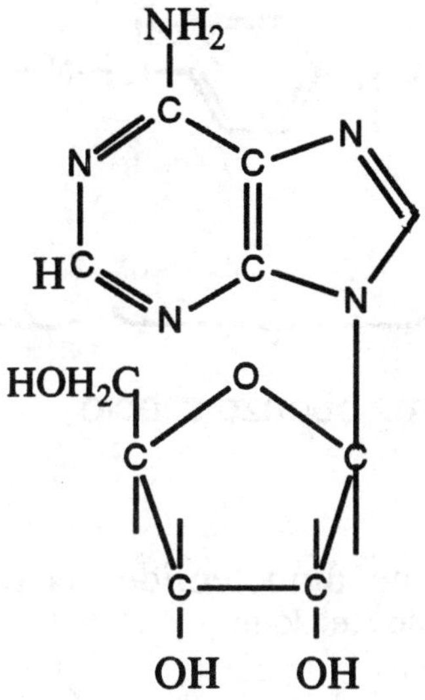

   Adenosine

   b. Acetylcholine

   $$H_3C-\overset{O}{\underset{\|}{C}}-OCH_2CH_2\overset{+}{N}(CH_3)_3$$

   c. Arachidonic acid
   CH$_3$(CH$_2$)$_4$CH=CHCH$_2$CH=CHCH$_2$CH=CHCH$_2$CH=CH(CH$_2$)$_3$COOH

   d. See next page.

Chapter 2 Water: The Solvent for Biochemical Reactions

Folic Acid

Pteridine derivative — p-Aminobenzoic acid — Glutamic acid

e. Nicotinamide adenine dinucleotide (NAD⁺, NADH)

Reduced form   Oxidized form

28

Chapter 2 Water: The Solvent for Biochemical Reactions

5. D-ribose and D-fructose are both monosaccharides, which are shown in Chapter 10. The open chain or linear configurations of D-ribose may be described as an <u>aldopentose</u>, while the D-fructose is a <u>ketohexose</u>. Point out how the underlined names help describe these molecules.

ANSWERS TO EXERCISES

1. a. a carbonyl group: aldehydes, ketones, carboxylic acids, esters (including lactones which are cyclic esters) and amides.
    b. a nitrogen atom: amines and amide.
    c. an -OH group: alcohols and carboxylic acids.

2. a.

    $H_3CCH_2CH_2COOH + H_3CCH_2CH_2OH \longrightarrow H_3CCH_2CH_2\overset{\overset{O}{\|}}{C} - O - CH_2CH_2CH_3 + H_2O$

    b.

    $HO-\overset{\overset{O}{\|}}{\underset{\underset{OH}{|}}{P}}-OH \; + \; H_3CCH_2OH \longrightarrow \; ^-O-\overset{\overset{O}{\|}}{\underset{\underset{O^-}{|}}{P}}-O-CH_2CH_3 \; + H_2O + 2H^+$

    c.

    ADP (structure shown in question) + H_3PO_4 ----->

29

Chapter 2 Water: The Solvent for Biochemical Reactions

$$\text{ATP structure} + H_2O + 2H^+$$

Adenosine triphosphate (ATP)

For convenience, the phosphoric acid reactant in reactions 2b and 2c is written as the completely protonated form in the reactants and the dissociated form in the products (the ester and the acid anhydride).

d.

$$\begin{array}{c}CH_2\text{-}OH\\|\\CH\text{-}OH\\|\\CH_2\text{-}OH\end{array} + \text{excess } R\text{-}\overset{O}{\underset{\|}{C}}\text{-}OH \longrightarrow \begin{array}{c}CH_2\text{-}O\text{-}\overset{O}{\underset{\|}{C}}\text{-}R\\|\\CH\text{-}O\text{-}\overset{O}{\underset{\|}{C}}\text{-}R\\|\\CH_2\text{-}O\text{-}\underset{\|}{C}\text{-}R\\\phantom{CH_2\text{-}O\text{-}}O\end{array} + 3H_2O$$

3.

a.

$$\begin{array}{c}NH_3^+\\|\\HC\text{-}COOH\\|\\\underset{\beta}{CH_2}\text{-}\underset{\delta}{CH_2}\text{-}\underset{\gamma}{CH_2}\text{-}\underset{\varepsilon}{CH_2}\text{-}NH_3^+\end{array}$$

The amino group on the side chain is on the epsilon (ε) carbon.

Chapter 2 Water: The Solvent for Biochemical Reactions

    b.   In the IUPAC nomenclature, the numbering of the carbon atoms start at the acid group. Therefore, the structure for 5-hydroxylysine is:

$$\underset{\text{Position} \rightarrow \quad 1 \quad 2 \quad\; 3 \quad\;\; 4 \quad\;\; 5 \quad\; 6}{\text{HOOC} - \overset{\overset{NH_3^+}{|}}{\text{CH}} - \text{CH}_2 - \text{CH}_2 - \overset{\overset{OH}{|}}{\text{CH}} - \text{CH}_2 - NH_3^+}$$

4.   a.   alkene, amines, alcohols, ether
       b.   ester, quaternary amine
       c.   alkene, carboxylic acid
       d.   amines, alcohol, amide, carboxylic acids, (aromatic character)
       e.   amines, alcohol, ether, alkenes, phosphate esters, acid (phosphate) anhydride

5.   The underlined descriptions indicate three points: (1) the molecule is a sugar (the suffix -ose), (2) it contains a ketone or an aldehyde (the prefix keto- or aldo-), and (3) it contains 5 (pent-) or 6 (hex-) carbons. Note that in a sugar with n carbon atoms, there are (n-1) alcohol groups, with one alcohol group on each carbon.

3

Amino Acids and Peptides

    This chapter presents the **amino acids**, with emphasis on their acid-base chemistry and the structure of their side chain groups. The only structural difference among each of the 20 amino acids is the side chain group, which determines the distinctive character of the amino acid. Although the side groups are routinely classified according to whether they are 1) nonpolar, 2) neutral polar, 3) acidic or 4) basic, additional characteristics include the size, shape, charge and the chemical reactivity of the individual groups. The amino acids are amphoteric in character and therefore their form and properties are dependent on the pH of the solution. The **zwitterion** is the form of the amino acid that has a net zero charge. The reaction of two amino acids to produce a dipeptide, linked by a **peptide bond**, exemplifies the condensation reaction common in the formation of larger peptides and proteins. The peptide bond, along with the carbonyl bond, exhibits partial double bond character and therefore this peptide unit is planar. The single bonds to the α-carbon atom exhibit various degrees of restricted rotation, depending on the character of the side chain group. All the amino acids, with the exception of glycine, have at least one chiral center and are optically active.

Chapter 3 Amino Acids and Peptides
A Key Figure Revisted

Figure 3.4 Structures of the amino acids. The 20 amino acids found in proteins are shown in their prominent forms at pH 7. The R groups are shown as ball and sticks. The amino acids within the dotted lines have carboxyl groups in their side chains, either as free carboxyls or as amides.

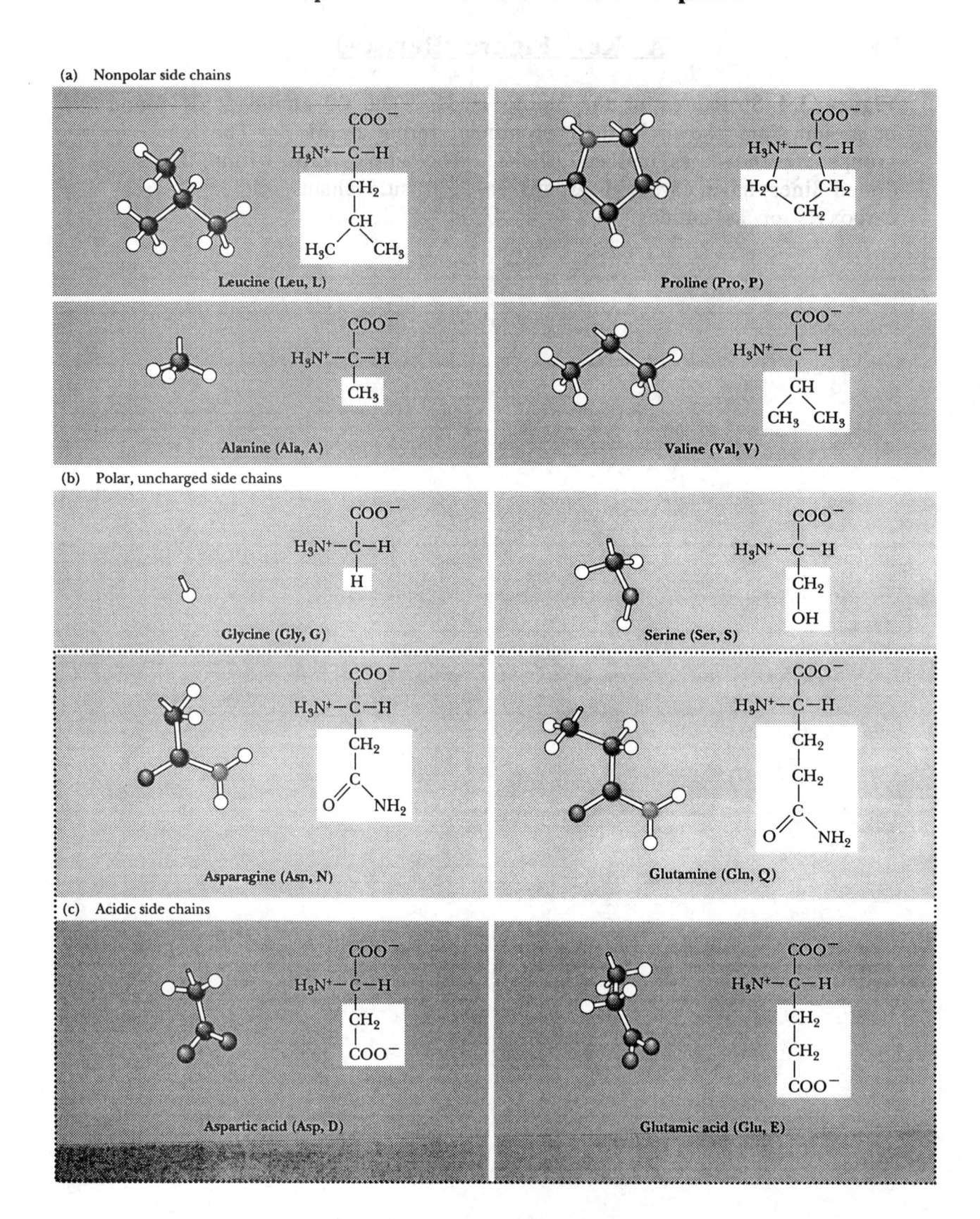

Chapter 3 Amino Acids and Peptides

(a) Nonpolar side chains

Leucine (Leu, L)

Proline (Pro, P)

Alanine (Ala, A)

Valine (Val, V)

(b) Polar, uncharged side chains

Glycine (Gly, G)

Serine (Ser, S)

Asparagine (Asn, N)

Glutamine (Gln, Q)

(c) Acidic side chains

Aspartic acid (Asp, D)

Glutamic acid (Glu, E)

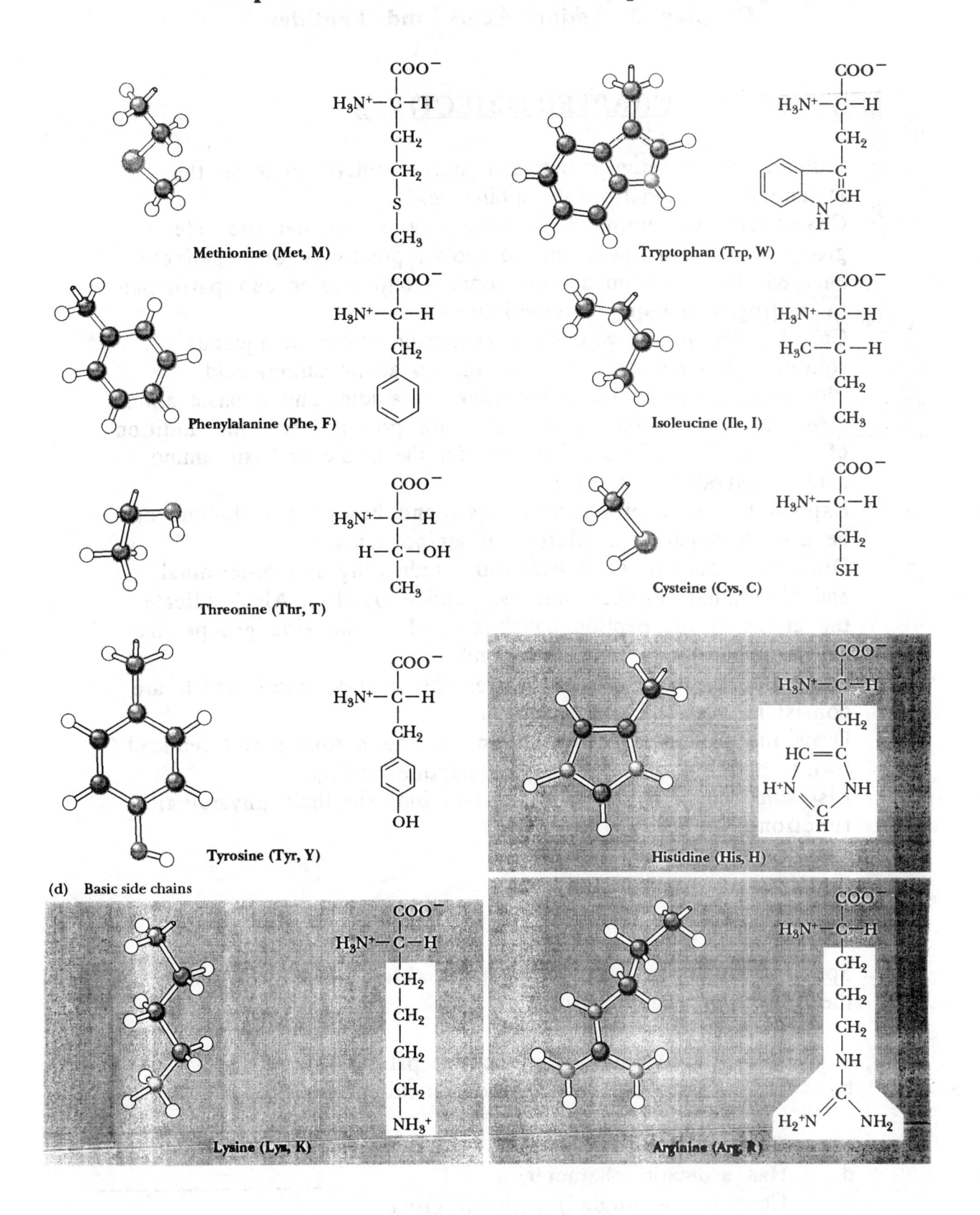

Chapter 3 Amino Acids and Peptides

Methionine (Met, M)

Tryptophan (Trp, W)

Phenylalanine (Phe, F)

Isoleucine (Ile, I)

Threonine (Thr, T)

Cysteine (Cys, C)

Tyrosine (Tyr, Y)

Histidine (His, H)

(d) Basic side chains

Lysine (Lys, K)

Arginine (Arg, R)

35

Chapter 3 Amino Acids and Peptides

CHAPTER OBJECTIVES

1. Indicate the difference between stereoisomers such as the D- and the L-isomers of an amino acid.
2. Characterize the amino acids according to whether the side group is aromatic, polar or non-polar, positively or negatively charged, has a chemical functional group and/or can participate in hydrogen bonding interactions.
3. Calculate the pH at which the zwitterion exists in aqueous solution for a nonpolar, a basic and an acidic amino acid.
4. Plot titration curves for a nonpolar, an acidic, and a basic amino acid and indicate the species that are present after the addition of 0.5, 1.0, 1.5, 2.0 (2.5, 3.0, also for the acidic or basic amino acids) equivalents of base.
5. Explain the basis of electrophoresis and how this technique can be used to separate a mixture of amino acids.
6. Draw the structure of a tripeptide, indicating the N-terminal and C-terminal residues and the peptide bonds. Also indicate the atoms in the peptide backbone and in the side groups that c an participate in a hydrogen bond.
7. Draw the resonance structures for the peptide bond which are consistent with its planarity.
8. Draw the structure for glutathione in the oxidized and reduced forms, pointing out the unusual peptide linkage.
9. List some small cyclic peptides and indicate their physiological functions.

EXERCISES

1. Specify the amino acid(s) that has a **side chain group** with the following characteristics:

    a. Is positively charged at neutral pH. _____
    b. Has an alcohol (or phenolic) functional group. _____

    c. Contains a thiol. _____
    d. Has aromatic character. _____
    e. Contains an amide functional group. _____
    f. Can participate in hydrogen bonding. _____

Chapter 3 Amino Acids and Peptides

2. List the **amino acids** that exhibit the following characteristics:

   a. Is positively charged at pH 7. _____
   b. Is negatively charged at pH 7. _____
   c. Can form disulfide bonds. _____
   d. Is not chiral. _____
   e. Has more than one chiral carbon. _____

3. Draw the structure for the zwitterion of alanine and lysine and indicate the pH at which they would exist. The pK values for alanine are 2.34 and 9.69, while those for lysine are 2.18, 8.95 (α-NH$_3$) and 10.53 (side chain).

4. Three amino acids, gly, ser and trp, react to produce a tripeptide. Indicate all possible products in the reaction. Specify the amino and carboxyl-terminal residues in the products.

5. Glutathione is perhaps the most abundant peptide in living systems; however, this tripeptide has a very unusual structure. The descriptive name for glutathione is, γ-glutamyl-L-cysteinylglycine. Draw the tripeptide and indicate how the nomenclature for naming this tripeptide defines its structure.

6. Draw the conjugate acid and conjugate base pair for aspartic acid that exists at the following pH values: a) pH = 2.09; b) pH = 3.86; and c) pH = 9.82.

   Hint: These pH values correspond to the 3 pK$_a$ values for aspartic acid. There is a different acid and base conjugate pair at each of these points.

7. Consider the titration of histidine. What species exist in solution after 0.5, 1.0 and 2.5 equivalents of NaOH have been added? Briefly justify your answer.

8. Consider the following two peptides:

   i. arg-asp-cys-his-lys   and   ii. ala-ile-phe-trp-met

Chapter 3 Amino Acids and Peptides

   a. Which peptide would exhibit the simpler titration curve?
   b. How many groups can be titrated in each peptide?
   c. Which peptide can be oxidized to produce a cystine linkage?
   d. Which peptide would be the more soluble in water?
   e. The zwitterionic form of which peptide would occur at the lower pH value?

9. Electrophoresis can be used to separate molecules of different charge. An electric field can be set up {(+) ----------(-)} in a stationary matrix material. If the amino acids were placed in the middle of the material, neutral species would not migrate from this position; however, the positively charged species would migrate toward the negative electrode, while the negatively charged species would travel in the opposite direction toward the positive electrode. If a drop of solution at pH 5, containing asp, his, ala and lys, were placed in the middle of the material, indicate how the electric field would effect the position of each of the amino acids.

10. The structure for morphine is shown below, together with the formula for the pentapeptide, methionine enkephalin. Morphine has a potent physiological activity which mimics that of the enkephalin. Can you suggest a possible reason for this property?

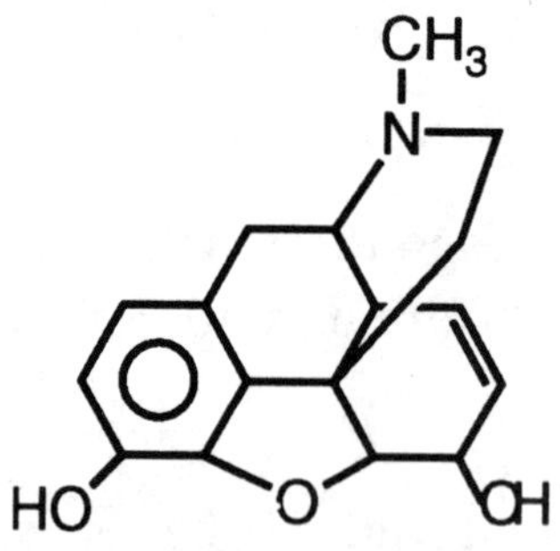

Morphine

Methionine enkephalin is: tyr-gly-gly-phe-met

Chapter 3 Amino Acids and Peptides

11. The idealized titration curve for glutamic acid is drawn below. Which species exist at points A-G on the diagram?

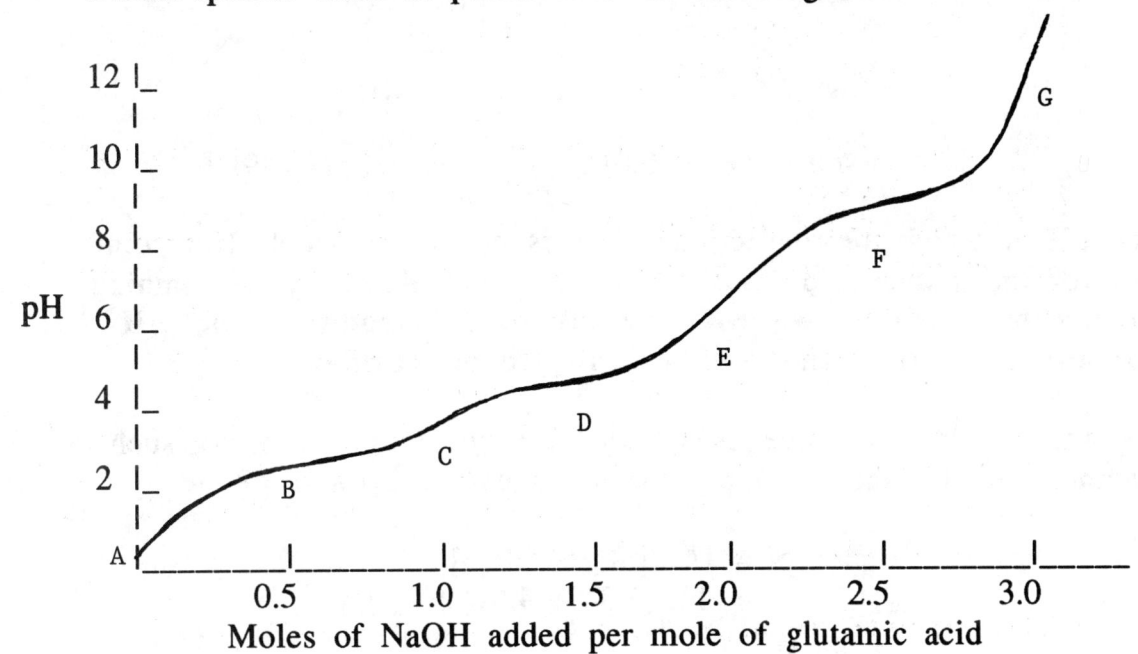

ANSWERS TO EXERCISES

1. a. lys, arg, his (approximately 10%)
    b. ser, thr, (tyr)
    c. cys
    d. phe, tyr, trp
    e. gln, asn
    f. arg, lys, his, glu, asp, asn, gln, ser, thr, tyr, trp

2. a. arg, lys, his (approximately 10%)
    b. asp, glu
    c. cys
    d. gly
    e. thr, ile

Chapter 3 Amino Acids and Peptides

3.

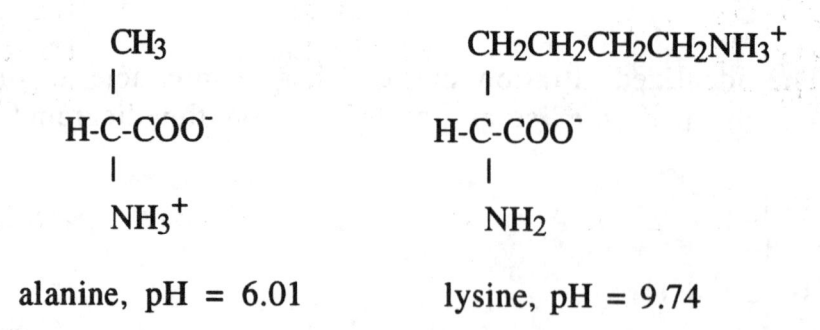

$$\begin{array}{cc} CH_3 & CH_2CH_2CH_2CH_2NH_3^+ \\ | & | \\ H\text{-}C\text{-}COO^- & H\text{-}C\text{-}COO^- \\ | & | \\ NH_3^+ & NH_2 \end{array}$$

alanine, pH = 6.01   lysine, pH = 9.74

The pH at which the zwitterion exists is called the isoelectric point (pI) for the amino acid. This pH value is calculated by determining the **average of the two pK_a values that "straddle" the pH domain in which the zwitterionic form resides.**

For amino acids with side groups that are not acidic or basic, such as alanine, the pI value is simply the average of pK_{a1} and pK_{a2}.

$$pI = 1/2\ [pK_{a1} + pK_{a2}]$$
$$= 1/2\ [2.34 + 9.69] = \underline{6.01}$$

For amino acids that have either an acidic or basic side chain group, the pI value is again the average of two appropriate pK_a values, one of which is always the pK_R, the pK for the side group. For lysine, a basic amino acid, pK_{a2}, is the other pK_a value which straddles the pH domain of the zwitterion. For acidic amino acids, such as aspartic acid and glutamic acid, pK_{a1}, is used.

$$pI = 1/2\ [pK_R + pK_{a2}]$$
$$= 1/2\ [8.95 + 10.53] = \underline{9.74}$$

4. The tripeptide products of this reaction will include all sequence combinations of the three amino acid residues. Therefore, there will be 9 different products. These products are listed below, with the amino terminal residue written to the left and the carboxyl terminal residue on the right.

Chapter 3 Amino Acids and Peptides

gly-gly-gly	ser-ser-ser	trp-trp-trp
gly-trp-ser	ser-gly-trp	trp-ser-gly
gly-ser-trp	ser-trp-gly	trp-gly-ser

$$^+H_3N\text{----------------}CO_2^-$$
(N-terminal residue)   (C-terminal residue)

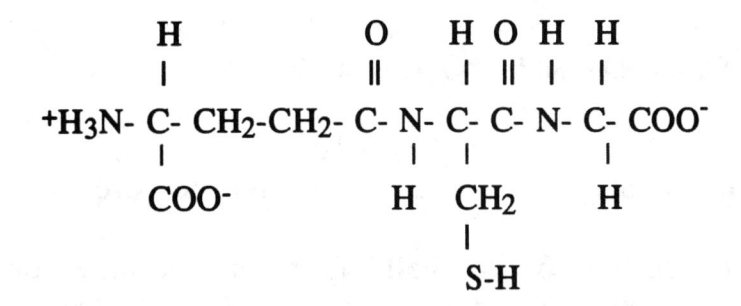

5.   γ-glu-cys-gly

```
          H           O   H O H   H
          |           ||  | || |   |
   +H3N- C- CH2-CH2- C- N- C- C- N- C- COO-
          |               |   |       |
         COO-             H  CH2      H
                              |
                             S-H
```

There are two peptide bonds in glutathione. The peptide bond between the cysteine and glycine residues is not unusual. The second peptide bond is unusual in that it is between the γ-carboxyl (the side group) of the glutamic acid residue and the amino end of the cysteine residue.

6.
 a. The species at pH = 2.09 are:

```
         H                          H
         |                          |
  HOOC-C-CH2COOH    and    -OOC-C-CH2COOH
         |                          |
        NH3+                       NH3+
                              (zwitterion form)
  Conjugate acid            Conjugate base
```

Chapter 3 Amino Acids and Peptides

b. The species at pH = 3.86 are:

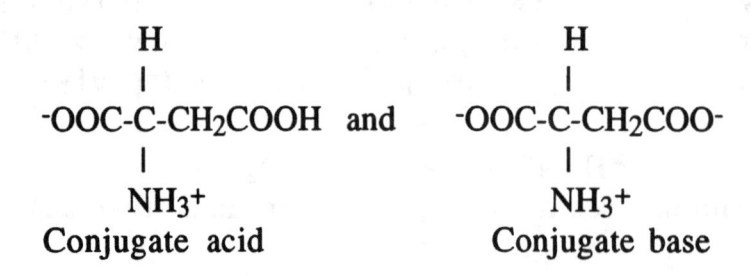

$$\begin{array}{c} H \\ | \\ {}^{-}OOC-C-CH_2COOH \\ | \\ NH_3{}^+ \end{array} \quad \text{and} \quad \begin{array}{c} H \\ | \\ {}^{-}OOC-C-CH_2COO^{-} \\ | \\ NH_3{}^+ \end{array}$$

Conjugate acid   Conjugate base

c. The species at pH = 9.82 are:

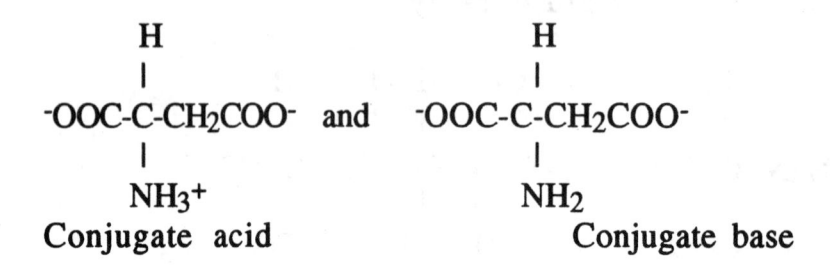

$$\begin{array}{c} H \\ | \\ {}^{-}OOC-C-CH_2COO^{-} \\ | \\ NH_3{}^+ \end{array} \quad \text{and} \quad \begin{array}{c} H \\ | \\ {}^{-}OOC-C-CH_2COO^{-} \\ | \\ NH_2 \end{array}$$

Conjugate acid   Conjugate base

7. a. After the addition of one-half equivalent of base, one-half of the carboxylic acid will be neutralized. This is the point at which equal amounts of the acid and the conjugate base are present {his(COOH) = his(COO⁻)}.

   After the addition of one equivalent of base, all of the protons from the acid group (-COOH) have been titrated.

   After two and one-half equivalents have been added, all the protons in the acid group (one equivalent) and the protons in the side chain (hisH⁺) group (the second equivalent) have been neutralized. In addition, half of the protons from the amino group (-NH₃⁺) have been neutralized.

8. a. Peptide (ii) has the simpler titration curve since there are no acidic or basic amino acid residues. The titration involves only the amino- and carboxyl-terminal groups.
   b. Peptide (i) has 7 groups that can be titrated. Every amino acid in the pentapeptide has a side chain group with a titratable hydrogen, in addition to the terminal groups. Peptide (ii) has only the terminal groups, which can be titrated by base.

42

Chapter 3 Amino Acids and Peptides

    c.   Peptide (i) only. It has a cysteine in it. Two molecules of this peptide can form a cystine bond.

    d.   Peptide (i) will be more soluble. Even though both peptides are about the same molecular weight, peptide (i) has many charged groups to interact with water. Peptide (ii) contains only hydrophobic and aromatic residues and will not interact with water in a positive way to manifest solubility.

    e.   The zwitterion form of peptide (ii) will have a lower pH since it contains no residues which have basic side chain groups.

9.   The given amino acids exist in the following forms at a pH of 5 The direction of migration indicated.

asp;	negatively charged;	will migrate toward + end
his;	positively charged;	will migrate toward - end
ala;	neutral;	will not migrate
lys;	positively charged;	will migrate toward - end

10.

Morphine   tyr-gly-gly-phe-met

Tyrosine, the amino-terminal residue of the met enkephalin, is structurally very similar to that in the upper, left-hand part of the morphine structure. It is thought that both molecules interact in a similar way with the same cellular target.

11.   Point A;   100% completely protonated form; species with a (+1) charge
        Point B;   50% of the species with (+1) charge and 50% zwitterion
        Point C;   100% zwitterion species
        Point D;   50% zwitterion species and 50% of the species

Chapter 3 Amino Acids and Peptides

              with (-1) charge
Point E;   100% of the species with (-1) charge
Point F;   50% of the species with (-1) charge and 50% of species with (-2) charge
Point G;   100% of the species with (-2) charge

The Three-Dimensional Structure of Proteins

4

Proteins are linear polymers of amino acid residues, which participate in a wide variety of functions, including escorting small molecules between cells, guarding species from invasion by foreign organisms, and playing an instrumental role in sexual development. The unique and complex structure of each protein is determined by the amino acid sequence of the protein. This intricate three-dimensional structural arrangement in space in governed by weak, non-covalent forces between the amino acid residues. Secondary structures, such as α-**helices** and β-**pleated sheets**, are held together by hydrogen-bonding interactions. The clustering of these secondary-structural units often forms a supersecondary structure. The unique tertiary structure of a protein is produced by hydrophobic and electrostatic interactions, in addition to hydrogen bonding. Proteins with more than one subunit exhibit an additional level of structure, a quaternary structure, which results from interactions involving any or all of the non-covalent forces. The native structure of a protein is almost always the biologically active form. Native

Chapter 4 The Three-Dimensional Structure of Proteins

proteins can be converted to inactive, denatured forms by chemicals or reagents that disrupt non-covalent forces. **Myoglobin** and **hemoglobin** are metalloproteins that contain a heme prosthetic group in addition to the protein segment. Since both proteins have been extensively characterized both structurally and chemically, they serve as comprehensive models for protein structure and show how the interactions of small molecules can influence protein conformation. Myoglobin, an oxygen storage protein, is a prototypic protein that simply binds to this small molecule (O_2). On the other hand, the four subunit protein, hemoglobin, represents an allosteric O_2 carrier, in which O_2 acts as a homotropic effector, producing a cooperative binding profile. The O_2 binding profile is further influenced by the binding of heterotropic effectors such as H^+, CO_2 and 2,3-bisphosphoglycerate. The influence that H^+ and CO_2 exert on O_2 binding is referred to as the **Bohr effect**. Finally the character of muscle cells is presented and the roles of myosin, actin and ATP in the mechanism of muscle contraction are outlined.

Interchapter A presents a variety of chromatographic and electrophoretic techniques that are routinely utilized to fractionate, purify and characterize amino acids and proteins. These physical techniques take advantage of differential solubility, the charge of the molecule, or the selective affinity for another molecule, as a means to separate proteins. The amino acid composition of a protein is routinely obtained by **ion-exchange chromatography** after the chemical hydrolysis of the protein. Procedures to determine the primary structure (that is, the sequence) of a protein use chemical reagents and enzymes that cleave at specific sites within the protein. These smaller polypeptide chains can be separated and then sequenced using the **Edman degradation** procedure.

Chapter 4 The Three-Dimensional Structure of Proteins
Some Key Figures Revisited

Figure 4.3a The α-helix. From left to right, ball-and-stick model of the α-helix, showing terminology; ball-and-stick model with plain or peptide groups shaded; computer-generated space-filling model of the α-helix; outline of the α-helix.

47

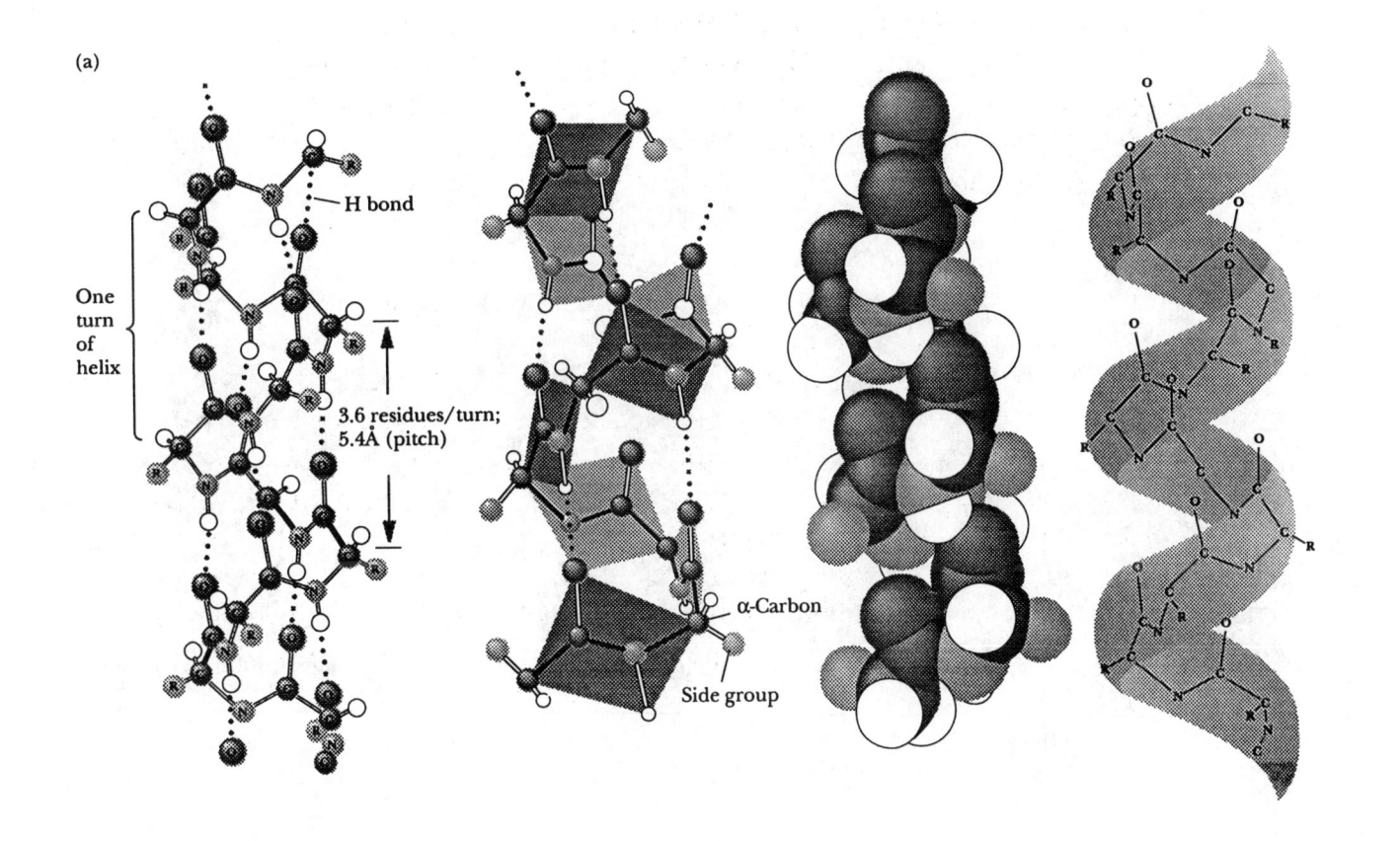

Chapter 4 The Three-Dimensional Structure of Proteins

Figure 4.4 ß-pleated sheet structures. (a) In an antiparallel pleated sheet, the peptide chains run in opposite directions from the N-terminal to the C-terminal ends. (b) In a parallel pleated sheet, the peptide chains run in the same direction.

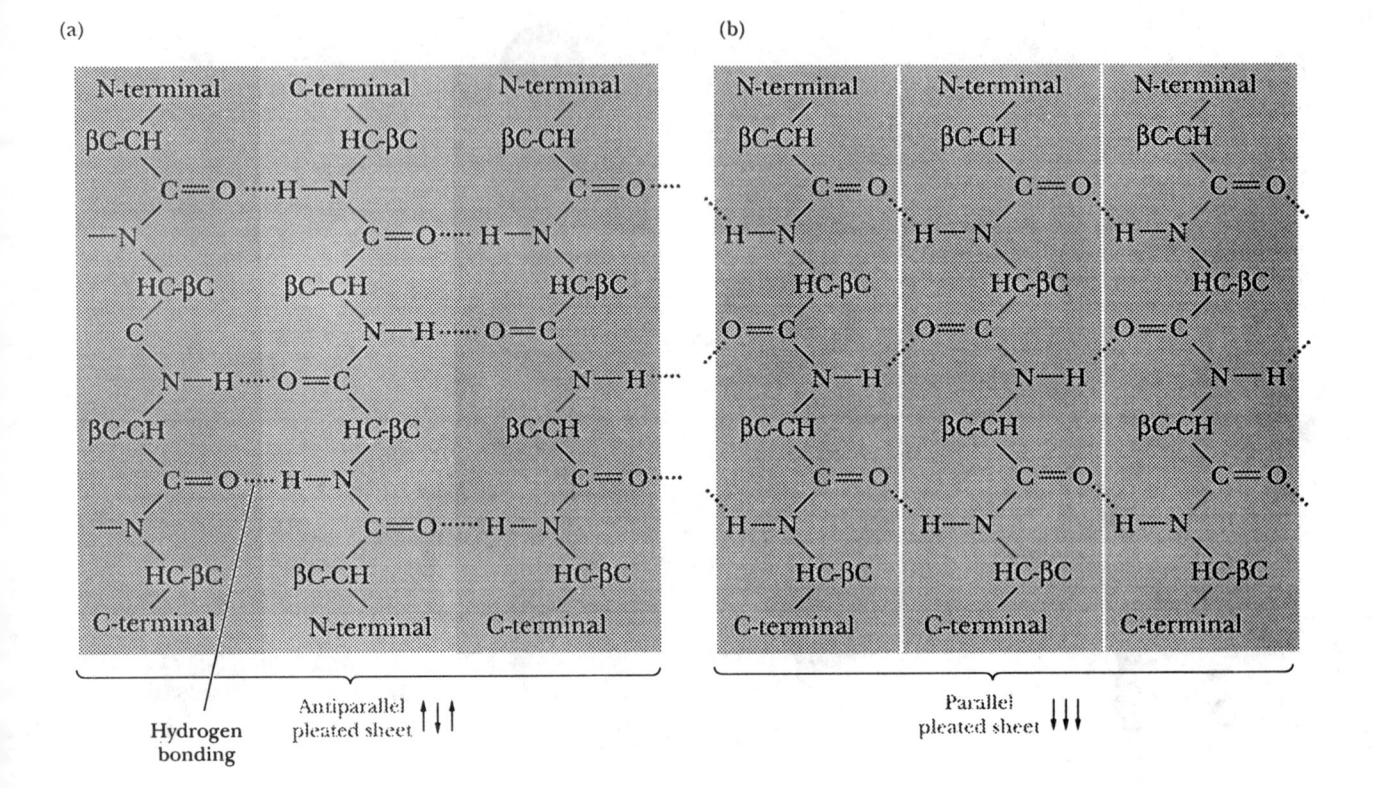

Chapter 4 The Three-Dimensional Structure of Proteins

CHAPTER OBJECTIVE

1. Draw the structure for a dipeptide and point out the peptide bond, the C-terminal and the N-terminal residues.
2. Name and describe the four levels of protein structure, indicating the types of forces associated with each level of structure.
3. List the four fundamental bonding interactions found in all proteins and the energy associated with each.
4. Describe the difference between a structural domain and a subunit.
5. Explain specifically how extreme pH, detergents, heat, high salt concentration or the addition of a reagent, such as urea, acts to denature proteins.
6. List the characteristics of an α-helix and the interactions that stabilize it.
7. Characterize the structure and the bonding in a β-pleated sheet and indicate the difference between a parallel and an anti-parallel β-pleated sheet.
8. Describe the characteristics of a supersecondary structure and what specifically constitutes a β-barrel.
9. Characterize the structure and bonding in a collagen helix and describe how it differs from an α-helix.
10. Describe the structure of myoglobin and explain the role of the prosthetic group in its function.
11. Discuss the O_2 binding curves for myoglobin and hemoglobin. What do these curves say about the nature of the binding of O_2 in each metalloprotein?
12. Indicate how H^+, CO_2 and 2,3-bisphosphoglycerate influence the binding of O_2 in hemoglobin.
13. Define the gross structural characteristics of a myofibril.
14. Outline the mechanism of muscle contraction.

Interchapter A:
Experimental Mehods for Determining Protein Structure

1. Describe the characteristic features of chromatography and the physical basis for the separation of proteins by HPLC and ion-exchange chromatography.
2. Outline the basis for separating amino acids or peptides by ion exchange chromatography and by electrophoresis, pointing out the unique features of SDS-PAGE.

Chapter 4 The Three-Dimensional Structure of Proteins

3. Describe the unique features involved in separating molecules by molecular sieve and by affinity chromatography.
4. Indicate the experimental methods used to determine the N-terminal and the C-terminal amino acid residues in a protein.
5. Indicate the chemical reagents and enzymes used to produce sequence specific cleavage of proteins and describe their specificity.
6. Outline the reagents and the strategy used in the Edman procedure to sequence peptides.

EXERCISES

1. How is the native structure of a protein or enzyme defined?

2. What are the forces that take part in determining the following level of protein structure?

    a. Primary
    b. Secondary (α-helix and β-pleated sheet)
    c. Tertiary
    d. Quaternary

3. Write the equation for the reaction of β-mercaptoethanol with a protein.

4. Explain how the following reagents will specifically denature proteins.

    a. Urea
    b. Lowering the pH to 2
    c. Increasing the salt concentration from physiological concentration (0.15 M NaCl) to 1.0 M NaCl.

5. Write the products formed in the reaction of the reagent or enzyme listed below with the pentapeptide, ala-trp-met-his-lys. Show the structure for the modified amino acid, if one is produced.

    a. Phenylisothiocyanate
    b. BrCN
    c. Trypsin
    d. Chymotrypsin

Chapter 4 The Three-Dimensional Structure of Proteins

6. A protein of 200 amino acid residues occurs naturally in a completely α-helical arrangement. By altering solution conditions, the protein is converted to a parallel β–sheet arrangement in which one molecule interacts side-by-side with another.

    a. What is the length of the α-helix and the β-pleated sheet?
    b. In the conversion of the α-helix to the β-pleated sheet, is the length of the protein increased or decreased?

7. Upon the addition of O_2 to a solution of either myoglobin (Mb) or hemoglobin (Hb), the Fe^{+2} is said to undergo **oxygenation** and not **oxidation**. Explain.

8. Consider the following experimental techniques used to separate a mixture of proteins.

    a. Affinity chromatography
    b. Electrophoresis
    c. Molecular-sieve chromatography
    d. Anion-exchange chromatography
    e. Isoelectric focusing
    f. SDS-PAGE

Indicate the physical basis for separating a mixture of proteins by each technique and indicate which protein would be the first to elute from the column or have the greatest migration in the gel?

9. X-ray crystallography has played an especially significant role in our understanding of the structural and mechanistic aspects of biomacromolecules.

    a. What particle in the atom is responsible for the scattering of the X-rays and producing the diffraction pattern?
    b. In a protein such as myoglobin, which has N, C, O, H, Fe, and S atoms, what is the relative intensity of the scattering from each of these atoms?

Chapter 4 The Three-Dimensional Structure of Proteins

10. Hemoglobin is a multisubunit protein which binds four molecules of O_2 in a cooperative manner.

    a. How is cooperative binding expressed in terms of the K (binding constants) and ΔG^o values for O_2 binding?
    b. Suggest a mechanism by which this cooperativity might occur.

11. After carrying out a number of separation procedures to isolate a pure enzyme, an investigator performed an enzymatic assay on a solution of the purified fraction and was elated to find that the fraction exhibited a high enzymatic activity. In order to be thorough and add further support to his feeling of finally achieving success, the investigator analyzed the protein fraction by SDS-PAGE and isoelectric focusing (IEF). The results are shown below. How do you interpret these data?

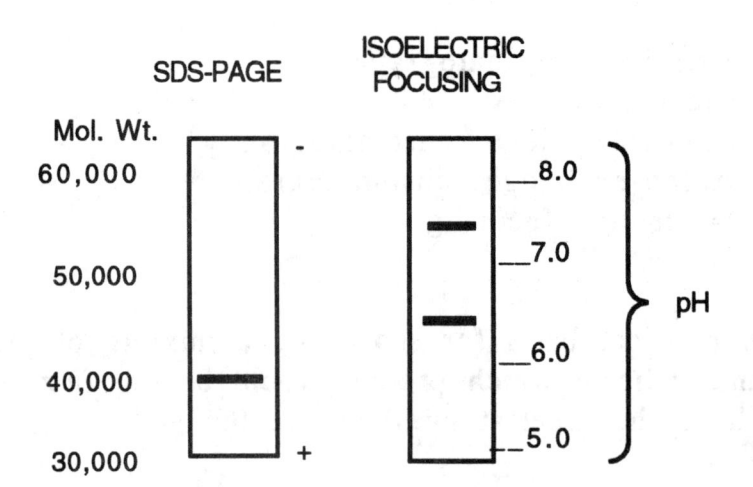

12. Indicate which of the following statements are true or false. If a statement is false, correct it.

    a. The heme group in each subunit of hemoglobin is covalently bound to the protein by the Fe^{+2}.
    b. Molecular oxygen binds to myoglobin in a non-cooperative manner and exhibits a sigmoidal binding curve.
    c. Decreasing the solution pH from pH 7.6 to 7.2 will increase the O_2 binding in hemoglobin.
    d. In the supersecondary structure referred to as βαβ, the β-pleated sheets are in a parallel arrangement.
    e. The Bohr effect refers to the influence that O_2 binding has on the binding of CO_2 in hemoglobin.

Chapter 4 The Three-Dimensional Structure of Proteins

    f.   The binding of H$^+$ to hemoglobin alters the quaternary structure of hemoglobin.

    g.   Electron microscopy permits the observation of the light and dark bands in the myofibrils of muscle.

    h.   Actin has ATPase activity.

13. Consider the composition and arrangement of these components in a muscle fiber (i.e., a muscle cell). Match the items in the right column with the corresponding terms in the left column.

    a.   Thin filaments   1.   Includes the region of thick filaments and region of overlapping thin and thick filaments.

    b.   Thick filaments   2.   Repeating units in myofibrils
    c.   H zone   3.   Myosin protein
    d.   Myofibrils   4.   Zones which contain only thick filaments.

    e.   An A band   5.   Actin, troponin and tropomyosin
    f.   Sarcomere   6.   Individual muscle fibers

14. The following Figure shows a schematic drawing of a skeletal myofibril. Using the list below, label the structures and the regions indicated.

    Myofibril   A band   I band

    Z line   H zone   Thin filament

    Sarcomere   M line   Thick filament

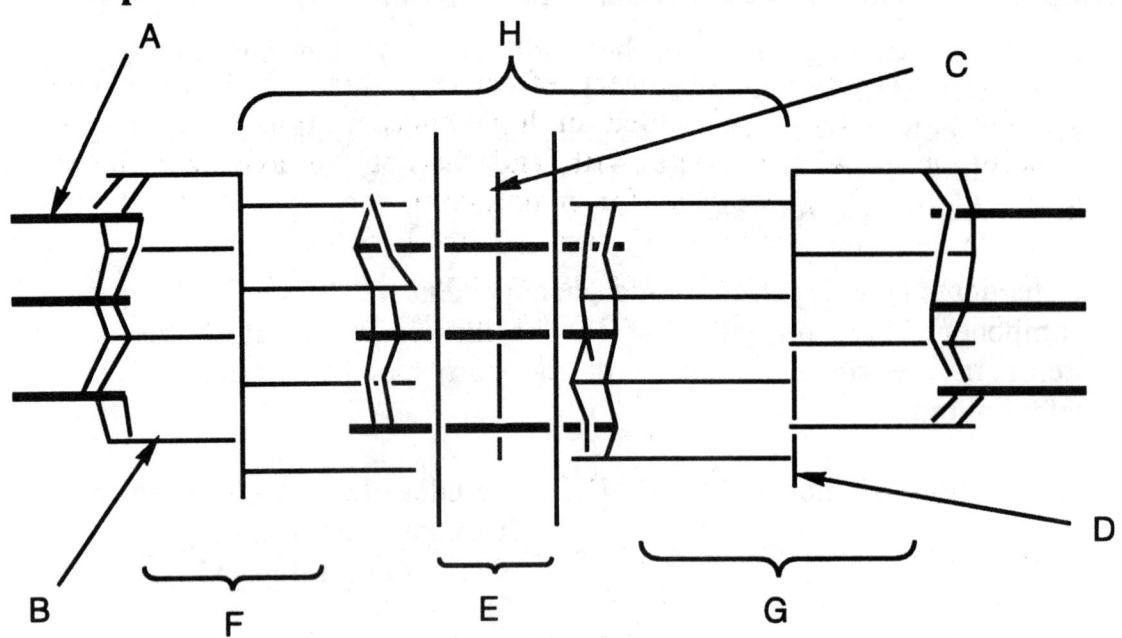

Chapter 4 The Three-Dimensional Structure of Proteins

ANSWERS TO EXERCISES

1. The native structure is defined as that structural arrangement which exhibits optimal biological activity.

2. a. Covalent (peptide) bonds connect each amino acid residue in the primary structure.

   b. Hydrogen bonds involving the atoms in the peptide backbone (not the side chains) are involved in secondary structure.

   c. and d. Virtually all the non-covalent forces are involved in the tertiary and quaternary structure of a protein. These include hydrogen bonds, electrostatic interactions and hydrophobic interactions.

3. β-Mercaptoethanol reacts with cystines in a protein to reduce them. This involves the breaking of the disulfide bond.

$$2\ HSCH_2CH_2OH\ +\ \begin{matrix}\textbf{Oxidized}\\ \text{Protein}\\ |\ \ |\\ S-S\end{matrix}\ \rightleftharpoons\ \begin{matrix}\textbf{Reduced}\\ \text{Protein}\\ |\ \ |\\ HS\ \ SH\end{matrix}\ +\ HOCH_2CH_2S\text{-}SCH_2CH_2OH$$

54

Chapter 4 The Three-Dimensional Structure of Proteins

4. a. Urea is a molecule that very effectively forms hydrogen bonds in solution; therefore, if added in high concentration to a protein solution, it will compete with and disrupt the hydrogen bonds that are (in part) responsible for maintaining the protein in its native structure. The structure of urea is shown below. The oxygen and the nitrogen atoms can act as hydrogen-bond acceptors, while the hydrogens act as hydrogen-bond donors.

$$H_2N - \overset{\overset{O}{\parallel}}{C} - NH_2$$

   b. A solution at pH 2 will be very acidic and protonate many groups, such as glutamic acid and aspartic acid. Increasing the pH will change the character of the carboxyl groups and disrupt electrostatic interactions, which may have been structurally important under normal pH conditions.

   c. Increasing the salt concentration will disrupt the electrostatic interactions in the protein.

5. a. Phenylisothiocyanate reacts with the amino-terminal residue of the pentapeptide. Upon hydrolysis, a tetrapeptide and the modified N-terminal amino acid, alanine, in the form of a phenylthiohydantoin alanine, will be produced.

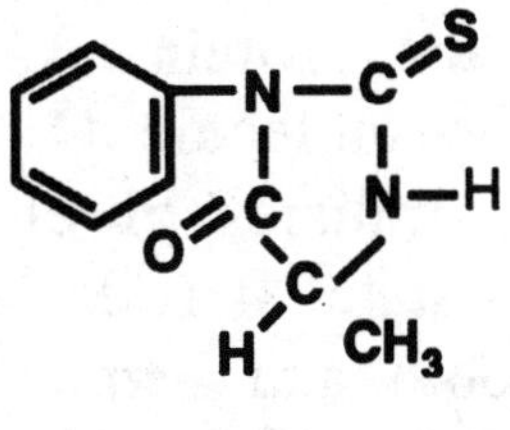

   [phenylthiohydantoin alanine structure]   + trp-met-his-lys

   b. BrCN reacts with the methionine residue, cleaves the peptide bond on the carboxyl side, and produces a tripeptide and dipeptide. The methionine residue is modified to a homoserine lactone (cyclic ester) in the process.

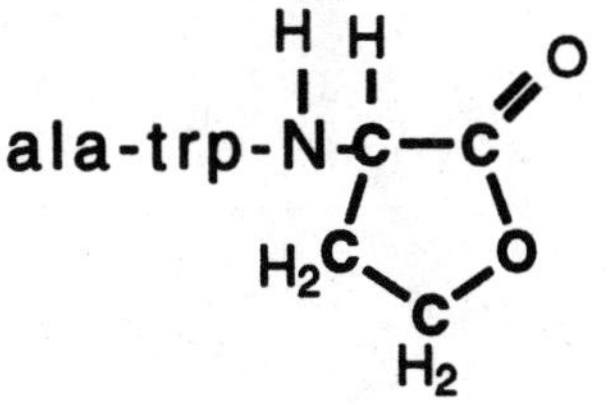

   ala-trp-N-[homoserine lactone structure]   + his-lys

Chapter 4 The Three-Dimensional Structure of Proteins

    c.   Trypsin cleaves a peptide at the carboxyl side of a lysine or arginine residue. Since the lysine residue is the C-terminal residue in this peptide, trypsin will have no effect on it.

    d.   Chymotrypsin cleaves on the carboxyl side of an aromatic residue, such as tryptophan, therefore, it will yield a dipeptide and a tripeptide.

                ala-trp   +   met-his-lys

6.   a.   The axial distance between adjacent amino acid residues in an α-helix is 0.15 nm and that for a β-pleated sheet is 0.35 nm. Therefore, the lengths for the two types of secondary structure are,

        α-helix: (199) x (0.15 nm) = 29.8 nm
        β-pleated sheet: (199) x (0.35 nm) = 69.6 nm

    b.   The α-helix is a very compact form of secondary structure, while the β-pleated sheet arrangement is much more extended. The structural conversion results in an increase in the length of the protein by more than a factor of two.

7.   The iron in both myoglobin and hemoglobin is in the plus two oxidation state, Fe^{+2}. It is essential for the functioning of both proteins that the iron remain in the ferrous state; this is accomplished primarily by the way in which the heme group is positioned within the architecture of the protein. As a result, oxygenation occurs (the binding of the molecule O_2), instead of oxidation in which there would be a transfer of electrons, resulting in the conversion of Fe^{+2} to the Fe^{+3} state. It is known that if the iron is oxidized to Fe^{+3}, both myoglobin and hemoglobin do not bind O_2 and therefore the proteins lose their essential activity.

Oxidation: Fe^{+2} (ferrous) + O_2 ----> Fe^{+3} (ferric)

Reversible oxygenation: $Fe^{+2} + O_2 \rightleftharpoons Fe^{+2}\text{-}(O_2)$

Chapter 4 The Three-Dimensional Structure of Proteins

8. a. A molecule which binds very strongly and specifically to the protein of interest is covalently attached to a chemically-inert matrix (the column material). Only proteins that exhibit a high affinity for a specific substrate will bind. All the proteins that have little or no affinity for the substrate elute first, while the proteins that bind strongly and specifically to the bound substrate are retained. A typical example of this is a column in which an antibody to a specific protein (A) is covalently bound. If a mixture of proteins were loaded on the column, all proteins would readily elute from the column, with the exception of the protein A. Protein A could then be eluted from the column under (different) elution conditions, which would disrupt the interaction between the antibody and protein A.

   b. Proteins migrate according to their charge. Proteins with the greater charge exhibit a greater mobility. The sign of the charge determines the direction of migration, with the positively-charged species migrating toward the negative pole, and the negatively-charged species migrating toward the positive pole.

   c. Proteins separate according to size. Small proteins enter into the gel permeation matrix and therefore are eluted only after long elution times. The larger proteins are excluded from the entering the gel matrix beads and are eluted at early times. The largest proteins elute first.

   d. An anion-exchange column is positively charged and interacts electrostatically with charged species. The column binds anionic species; molecules having the greater number of negatively-charged groups exhibit the highest affinity. Species which are positively charged, or do not interact with the positively-charged groups on the column material, are the first to elute.

   e. Isoelectric focusing separates proteins according to their characteristic pI value (i.e., the pH at which the protein has a net zero charge). Experimentally, this is an electrophoretic (gel) separation in a continuous pH gradient. If the protein mixture is loaded on the end of the gel that has a high pH, then the proteins with the lowest pI value exhibit the greatest migration before they attain the pH = pI.

Chapter 4 The Three-Dimensional Structure of Proteins

    f.   Proteins are separated by size under denaturing conditions in SDS-PAGE. The smallest protein will exhibit the greatest mobility.

9.   a.   X-rays are scattered by the electrons in the atom.

    b.   The scattering intensity is proportional to the number of electrons in the atom; therefore, the relative scattering intensity for the atoms in myoglobin would be: $Fe \gg S > O > N > C > H$.

10.   a.   Hemoglobin contains four subunits, each bearing a heme group. Each heme group contains an Fe^{+2}, which binds one molecule of O_2. The first O_2 has the most difficulty binding to Fe^{+2}. This means that its binding constant (K_1) is the smallest. The second O_2 binds more strongly (i.e., K_2 is greater than K_1). This trend continues until the four molecules of O_2 are bound. Since the successive K_n values increase, the corresponding ΔG^o values become progressively more negative.

    b.   Since the binding of the four O_2 molecules occurs on four different subunits, there must be a form of "molecular communication" between the subunits. The binding of the first O_2 alters the tertiary structure of the subunit to which it is bound. This change in the tertiary structure results in a change in the molecular contacts between the individual subunits. This effects a change in the tertiary structure of the second subunit, which enhances the binding of O_2 to this subunit. As a result, an overall change in the quaternary structure of the protein is produced. This trend continues until hemoglobin is saturated with four molecules of O_2.

11.   The SDS-PAGE data indicate that there is a protein of about 40,000 Daltons. Our friend is feeling confident and is ready to send out invitations for a party.
    But wait a minute! The IEF gel, however, reveals that there are two species, with individual pI values of 6.4 and 7.4.
These data indicate a potential caveat in SDS-PAGE. The **presence of a single band does not mean the presence of only one protein** in the sample. It could mean that there is (1) one protein in the sample; (2) multiple proteins with essentially the same molecular weight; or (3) one protein, but with a post-translational modification such as a phosphate group (or a number of other groups) added onto an amino acid residue. Such modifying groups usually have small molecular weights and add relatively little to

Chapter 4 The Three-Dimensional Structure of Proteins

the total molecular weight of the protein, such that the migration is indistinguishable from that of the unmodified protein. Since other interpretations are also possible, further experiments must be carried out before the investigator can claim success. This example emphasizes that before success is declared, a number of different properties of the protein must be examined and all results must be consistent with the presence of a single, unique protein.

12. a. True.
    b. False. A sigmoidal curve is a characteristic feature of cooperative binding in a protein.
    c. False. Decreasing the pH will decrease the extent of O_2 binding to hemoglobin.
    d. True.
    e. False. The Bohr effect refers to the influence of the binding of H^+ and CO_2 on O_2 binding in hemoglobin.
    f. True.
    g. True.
    h. False. Myosin is the ATPase essential for muscle contraction.

13. a. 5   b. 3   c. 4
    d. 6   e. 1   f. 2

14. A. Thick filament (myosin)   B.   Thin filament (actin)
    C. M line   D.   Z line
    E. H zone   F and G. Both are I bands
    H. Sarcomere

5

The Behavior of Proteins: Enzymes

For life to exist as we know it, biological reactions must proceed at enormous rates, with high specificity and with no side reactions. **Enzymes**, which are the catalysts in living organisms, serve this essential role. Substrates interact specifically with the enzyme in a region called the **active site** and are converted into products. The effect of substrate binding on the structure of the enzyme may be described by either the **lock-and-key** or the **induced-fit** models. The kinetic analysis of enzyme catalyzed reactions can be understood by the **Michealis-Menten** kinetic model, which assumes the intermediate formation of an enzyme-substrate complex. The enzymes are characterized by their K_M and V_{max} values, both of which can be obtained by plotting velocity as a function of substrate concentration in a **Lineweaver-Burk double reciprocal plot**. Molecules that reversibly inhibit enzyme activity can be experimentally classified as either **competitive** or **noncompetitive** **inhibitors**. **Allosteric** or **regulatory enzymes** have multiple subunits and exhibit more complex kinetics as a result of cooperative interactions and the effects from the binding of allosteric effectors. The kinetic profiles for these enzymes are not described by the Michealis-Menten model. In most cases, the

Chapter 5 The Behavior of Proteins: Enzymes

behavior of allosteric interactions can be described adequately by either the **concerted** or the **sequential** models. An alternate mechanism used to regulate enzyme activity involves the specific cleavage of covalent bonds the backbone in **zymogens** (an inactive precursor of an enzyme) to activate enzymatic activity in the product molecule. The mechanism or pathway by which an enzyme-catalyzed reaction takes place is directly linked to the amino acid residues in the active site. The architecture of the active site and the interactions between the amino acid residues and the substrate weaken or strain the appropriate substrate bond. In a wide variety of these reactions, a coenzyme directly participates in one or more steps in the conversion of substrate to product. A **coenzyme** is a non-protein substance that takes part in enzymatic reactions and is regenerated for further reactions.

Chapter 5 The Behavior of Proteins: Enzymes

Some Key Figures Revisited

Figure 5.14 The concerted model for allosteric behavior. The T (inactive) and R (active) forms of the enzyme are in equilibrium. The equilibrium lies to the left, in favor of the T form. The cooperative binding of substrate then shifts the equilibrium to the right, in favor of the R form.

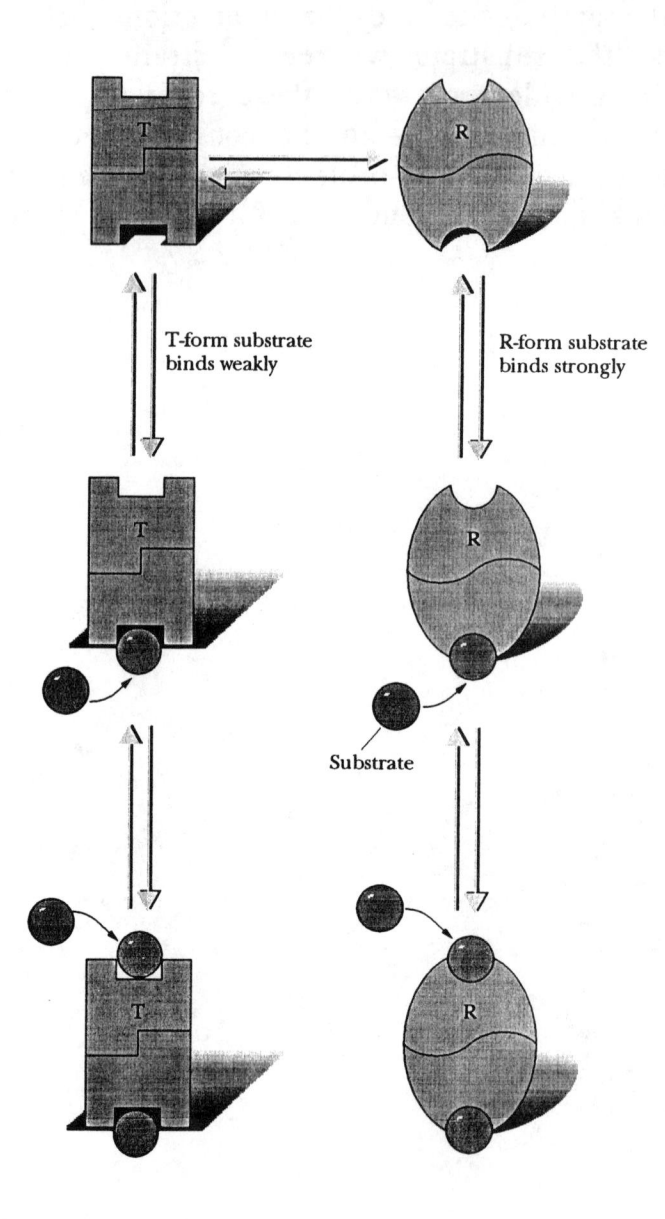

Chapter 5 The Behavior of Proteins: Enzymes

Figure 5.19 The mechanism of action of chymotrypsin. In the first stage of the reaction, the nucleophilic serine 195 attacks the carbonyl carbon of the substrate. In the second stage, water is the nucleophile that attacks the acyl-enzyme intermediate. Note the involvement of histidine 57 in both stages of the reaction. (From G. Hammes, 1982, *Enzyme Catalysis and Regulation*, Academic Press, New York)

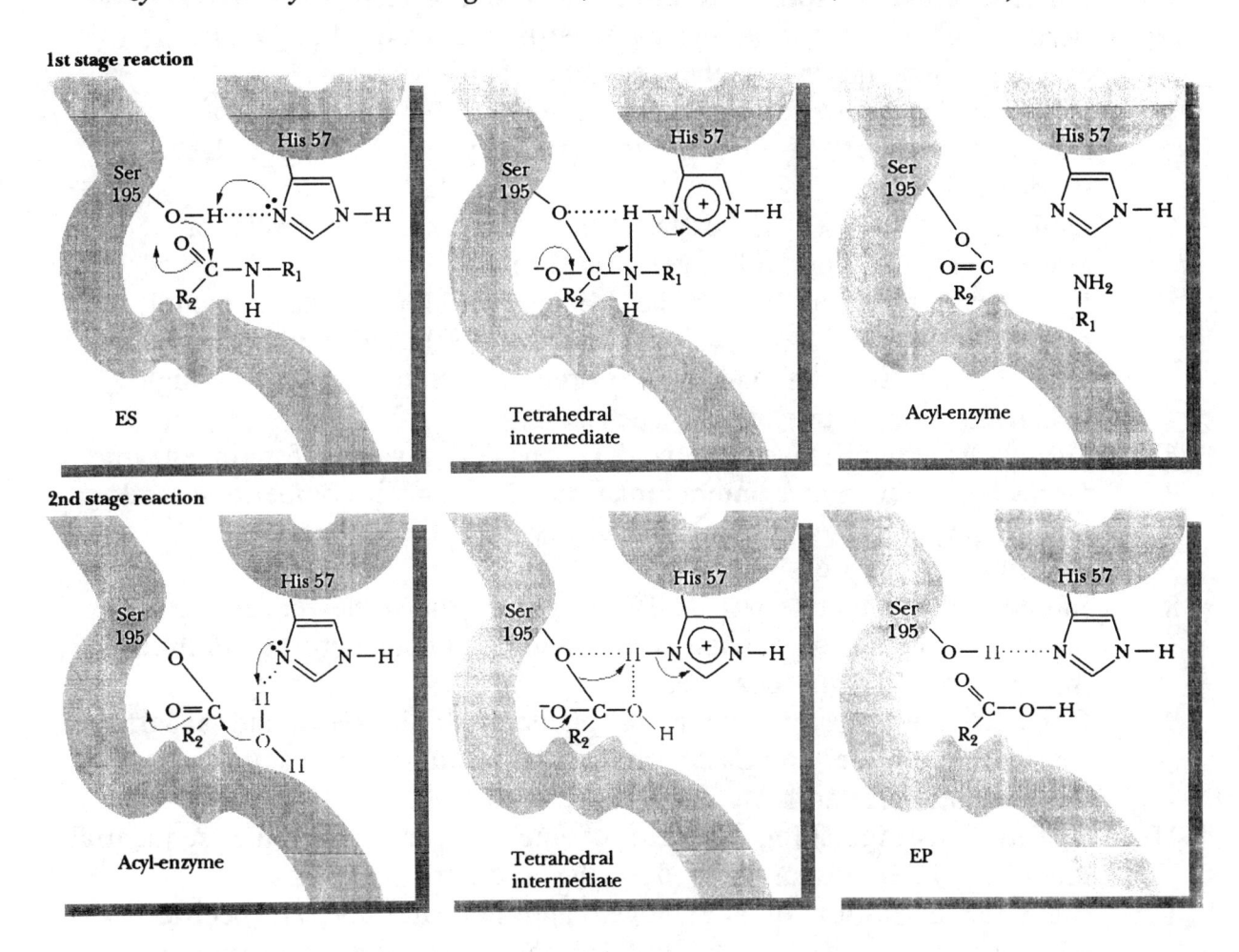

Chapter 5 The Behavior of Proteins: Enzymes

CHAPTER OBJECTIVES

1. Distinguish between the spontaneity and the rate of a reaction and indicate what factors determine each characteristic.
2. Indicate how a reaction mechanism differs from the overall equation that describes the reactants and products.
3. Describe the nature of the active site in an enzyme when a substrate interacts by 1) the lock-and-key and 2) the induced-fit model.
4. Outline the features of enzymes that display a hyperbolic or alternatively, a sigmoidal reaction profile.
5. For enzymes which follow the Michaelis-Menten kinetic model, derive the expression, $V = V_{max}[S]/(K_M+[S])$, and explain how this leads to a first-order and a zero-order reaction at low and high substrate concentrations, respectively.
6. Discuss the significance of the K_M and V_{max} values for an enzyme.
7. Describe the enzyme interactions involved with competitive and noncompetitive (reversible) inhibitors and the associated equilibrium expressions.
8. Indicate the characteristic differences in the Lineweaver-Burk plots as a result of a competitive and a noncompetitive inhibitor interacting with an enzyme.
9. Compare the allosteric behavior of hemoglobin (Hb) and aspartate transcarbamoylase (ATCase) and indicate the homotropic and heterotropic effectors in each case.
10. List the distinguishing features of the concerted and the sequential models, which are used to describe allosteric enzymes.
11. State the definition of a zymogen and indicate several specific enzymes in which this strategy for the regulation of enzymatic activity is used.
12. List the amino acid residues that play a direct role in the catalytic action of an enzyme.
13. Indicate the cofactors or amino acid residues in an enzyme that participate in general acid-base and Lewis acid-base catalysis and explain how they can specifically stimulate the reaction rate.
14. Name the characteristics of a coenzyme and describe how coenzymes participate in catalysis.

Chapter 5 The Behavior of Proteins: Enzymes

EXERCISES

1. What is the effect of a catalyst on the values of ΔG^o and $\Delta G^{o\ddagger}$?

2. The rate equation for a reaction is: Rate = $k [A]^1[B]^0$.
   a. Indicate the order of the reaction for A and B and the overall order of the reaction.

   b. If the reaction rate is determined to be 0.5 M/sec. when [A] = 0.01 M and [B] = 0.67 M, calculate the value of the rate constant.

3. Chymotrypsin has 28 serine residues; however, reaction with diisopropylphosphofluoridate produces only one covalently-modified serine residue. Explain.

4. Consider the following diagram, which shows the reaction profile for a catalyzed and uncatalyzed reaction. Label the coordinates of the plot. Also label the diagram appropriately with ΔG^o, $\Delta G^{o\ddagger}_{forward}$, and $\Delta G^{o\ddagger}_{reverse}$ for both the catalyzed and uncatalyzed reactions.

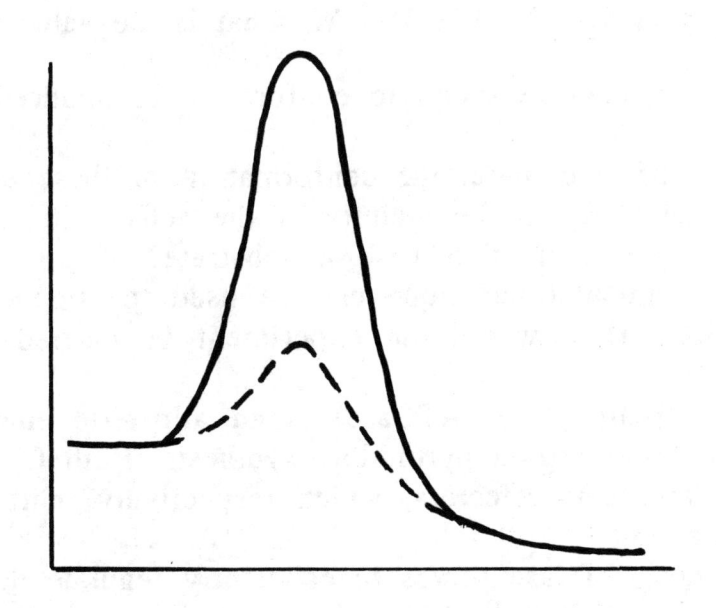

Chapter 5 The Behavior of Proteins: Enzymes

5. a. What fundamental parameters influence the value for the rate constant of an enzymatic reaction?

   b. What parameters influence the rate of an enzymatic reaction?

6. An enzyme with a maximum activity at pH 5.0 exhibits a greatly reduced activity at pH 7. Briefly explain.

7. Reactions in organic and inorganic chemistry typically proceed faster as the temperature is increased. This same trend is observed when an enzyme catalyzed reaction is carried out at 15, 25 and 40°C. However, at 45°C and 50°C, the reaction rate decreases sharply. Provide an explanation for this observation.

8. The value of K_M for an enzyme was determined to be 1×10^{-6} M. A competitive inhibitor is added to this enzyme-catalyzed reaction.

   a. How does the inhibitor affect the value of K_M and V_{max}?

   b. If the inhibitor concentration is 1×10^{-7} M and $K_{M apparent}$ (the K_M value determined in the presence of the inhibitor) was determined to be 1×10^{-3} M, what is the value for K_I?

9. A number of enzymes are found to conform to an induced-fit model.
   a. What can be said about the conformation of these enzymes, at or immediately in the vicinity of the active site, before and after the interaction with a substrate?
   b. What experimental technique can be used to support this proposition? How would the experiment be carried out?

10. Aspartate transcarbamoylase (ATCase) is an allosteric enzyme. It is involved in the first step in pyrimidine synthesis. Both CTP and ATP serve as allosteric effectors, which respectively inhibit and activate the enzyme.
    a. Explain how ATCase serves to efficiently regulate this pathway in nucleotide biosynthesis.

Chapter 5 The Behavior of Proteins: Enzymes

    b.   It is found that if *p*-hydroxymercuribenzoate reacts with ATCase, the allosteric character of the enzyme is lost; however, the substrates bind to the enzyme in a noncooperative fashion and the enzyme exhibits catalytic activity. Explain.

11.   It has been proposed that for enzymes which obey Michaelis-Menten kinetics, the K_M value for an enzyme may represent the approximate physiological substrate concentration in the cell. Examine a typical plot of V versus [S] and explain why this appears to be a reasonable suggestion.

12.   An inhibitor is discovered that interacts <u>only</u> with the enzyme-substrate complex, with no affinity for the free enzyme. This is actually the case for an <u>un</u>competitive inhibitor. Predict how this inhibitor influences the magnitude of the value of K_M and V_{max} for the enzyme.

13.   Indicate whether the following statements are true or false. If the statement is false, correct it.
    a.   The V_{max} value for an enzyme is independent of substrate and enzyme concentration.
    b.   Enzymes which conform to Michealis-Menten kinetics are not involved in any feedback regulation.
    c.   Enzymes reduce the value of ΔG^o more than that of $\Delta G^{\ddagger}$ for a reaction.
    d.   The formation of an intermediate, enzyme-substrate complex is a basic assumption in the Michealis-Menten model.
    e.   Substrates can be bound to the active site of an enzyme by either covalent or non-covalent interactions.
    f.   The K_M value for an enzymatic reaction is always an indication of how tightly the binding is between the substrate and the enzyme.
    g.   A molecule, such as diisopropylphosphofluoridate, which covalently binds within the active site of serine proteases, is an example of an effective competitive inhibitor.
    h.   The acid, HCl, can act as a general acid catalyst.
    i.   The clotting of blood provides an elegant example of how the activation of zymogens plays a critical part in a multistep process.
    j.   Allosteric enzymes always exhibit sigmoidal plots of V

Chapter 5 The Behavior of Proteins: Enzymes

versus [S].
k. The sequential model for an allosteric enzyme assumes that the R and the T forms of the enzyme are in equilibrium.
l. The greater the association constant, K_M, for the ES complex, the stronger the interactions between the enzyme and the substrate.

ANSWERS TO EXERCISES

1. A catalyst has no effect on the free-energy difference between the reactants and the product, ΔG^o. On the other hand, a catalyst decreases the value of $\Delta G^{o\ddagger}$, the activation energy required to bring the reactants to the transition state. Refer to Figure 5.1a in the text.

2. a. The reaction is first order for [A], zero order for [B] and first order overall.

   b. Rate = 0.5 M/sec. = k (0.01 M)^1(0.67 M)^0
              = 0.5 M/sec. = k (0.01 M)
      <u>k = 50 sec^{-1}</u>

3. This type of selective hyperreactivity is quite common in enzymes and indicates that the one serine is in a unique environment that makes it especially reactive relative to the others. The reactive residue is found in the active site.

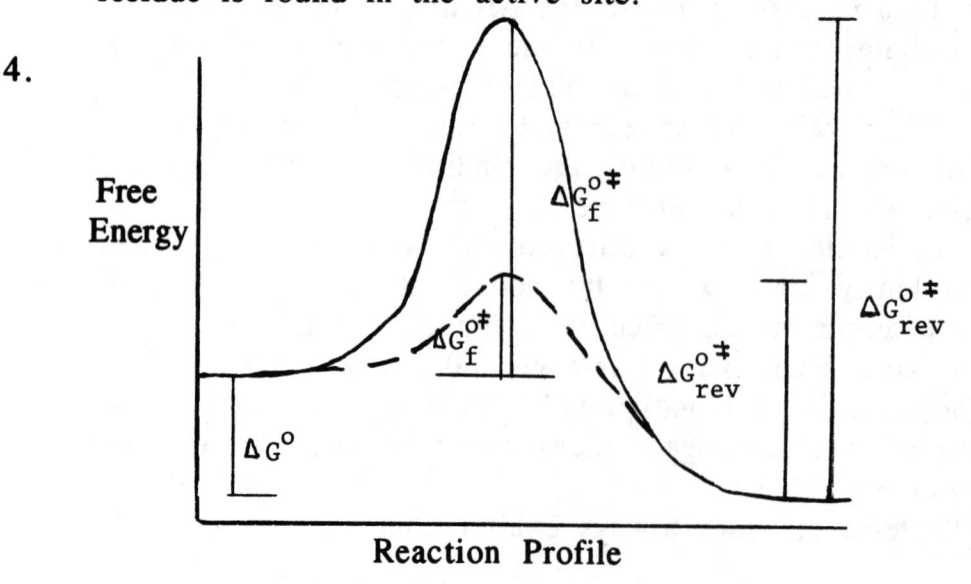

4.

Reaction Profile

Chapter 5 The Behavior of Proteins: Enzymes

5. a. The <u>rate constant</u> for an enzymatic reaction depends on the temperature and the $\Delta G^{o\ddagger}$ value.
   b. The rate of the reaction depends on the rate constant and the solution conditions, such as the concentration of substrate, pH, etc.

6. Enzymes are generally very sensitive to the pH of the solution. This is because the acidic or basic amino acid residues can become protonated or deprotonated as the pH changes. Since the character of the amino acids influences the native structure of the enzyme, and especially the nature of the active site, the catalytic activity is often strongly influenced by pH.

7. Enzymes are sensitive to the temperature of the solution because their secondary and tertiary (and if a multisubunit enzyme, the quaternary) structure is determined by a multitude of weak non-covalent interactions. At some point, as the temperature is increased, these forces will be sufficiently disrupted that the native structure is altered and the catalytic activity is reduced or completely lost.

8. a. V_{max} is unchanged by a competitive inhibitor, while the apparent K_M value is increased by the factor $(1 + [I]/K_I)$.

   b. $K_{M\ apparent} = K_M(1 + [I]/K_I)$
      $K_I = 1 \times 10^{-10}$ M

9. a. Enzymes which can be described by the induced fit model have a conformation which is not complementary to that of the substrate. The conformation will change in and about the active site as a result of the substrate binding.
   b. Isolate a crystal of the enzyme with and without bound substrate and determine the structure by X-ray crystallography. The structures would be expected to be different in and about the active site of the enzyme. In actual practice, it is not possible to obtain the structure of an enzyme interacting with a true, intact substrate; therefore, a competitive inhibitor or a modified substrate is used in place of the substrate.

Chapter 5 The Behavior of Proteins: Enzymes

10. a. CTP is the final product in this pathway. It acts as a feedback allosteric inhibitor as it binds to an effector site (different and distinct from the substrate site) to reduce the activity of ATCase. ATP, on the other hand, is not involved directly in this pathway and is an allosteric activator. Since ATP and CTP bind to the same effector site, they will compete with each other. The effector that binds will depend on the relative concentrations of each effector and their respective binding constants. The reason that ATP exerts an effect on this pathway is because both ATP and CTP are needed for RNA and DNA synthesis. The role of ATP is to aid in coordinating the production of pyrimidine nucleotides with that for purine nucleotides. It is even further complicated in that ATP is also required as the cellular source of energy.
    b. P-hydroxymercuribenzoate binds to ATCase to dissociate the enzyme into the individual subunits. The substrate can continue to bind to the active site and the enzyme exhibits activity, but it has lost its allosteric and regulatory character.

11. The V versus [S] plot is shown on the right for a simple enzyme obeying Michealis-Menten kinetics. The K_M value represents the substrate concentration at 1/2(V_{max}). The proposal suggests that if the [S] were significantly less than this value, most of the enzyme activity would be, in effect, unused and therefore, wasted. On the other hand, if the [S] in the cell were twice as large or much larger than the K_M value, the enzyme would be saturated and small changes in the [S] would not effect the rate of the reaction. With the [S] at approximately the value of K_M, much of the enzymatic activity is being used. If the concentration of substrate within the cell were to change, the rate of the catalyzed reaction would change accordingly, so that some degree of regulation would be effected by changing metabolic states.

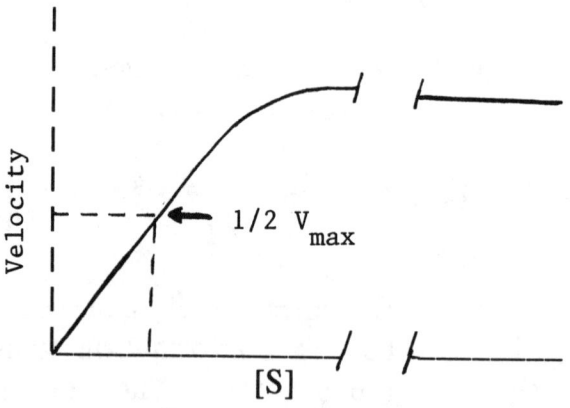

Chapter 5 The Behavior of Proteins: Enzymes

12. This type of inhibition decreases the value for both V_{max} and K_M.

13.
- a. False. V_{max} is not really a fundamental characteristic of an enzyme. It depends on the <u>conditions</u> which are used to carry out the reaction (i.e., the amount of enzyme and substrate).
- b. True.
- c. False. Enzymes do not affect the ΔG^o value.
- d. True.
- e. True.
- f. False. This is only true under conditions in which $k_{-1} > k_2$.
- g. False. Competitive inhibitors, and also noncompetitive inhibitors, must interact reversibly with the enzyme by noncovalent interactions.
- h. False. General acid catalysis requires the reversible donation and acceptance of a proton from a weak acid.
- i. True.
- j. True.
- k. False. Only the T form of the allosteric enzyme exists. The binding of an effector induces a conformational change from the T to the R form.
- l. False. K_M is a <u>dissociation</u> constant for the ES complex, as shown below. The smaller the K_M value, the stronger the binding is in the [ES] complex.

$$ES = E + S \qquad K_M = ([E][S])/[ES]$$

6

Nucleic Acids: How Structure Conveys Information

Deoxyribonucleic acid (DNA) is the vital polynucleotide which is the repository for the genetic information in all cellular life forms. It provides the molecular vehicle for the expression of this genetic information during the life of the cell and for the transmission of hereditary characteristics from one generation to the next. The secondary structure of DNA is not uniform and may be found in the **A**, **B** and **Z-forms**. The structure and function of DNA is influenced further by **supercoiling**, which creates different tertiary structural forms. The primary (order of the bases in a polynucleotide), secondary (the three-dimensional conformation of the backbone, e.g., the double helix) and tertiary (the supercoiling of the molecule) structural aspects of DNA provide the basis for understanding its essential role in the dynamic processes of replication, transcription and recombination. These important cellular transactions are presented in more detail in the following chapter. The three forms of cellular **ribonucleic acid** (RNA) are described

Chapter 6 Nucleic Acids: How Structure Conveys Information

and their roles in the translation process are outlined. An enormous number of enzymes act on DNA and RNA substrates. **Restriction endonucleases**, which play a pivotal role in our ability to sequence DNA (i.e., determine its primary structure), and in recombinant DNA technology, are highlighted. **Genetic engineering** has already led to genetically-altered agricultural crops, which exhibit increased resistance to pests and severe weather. Transgenic animals, which contain foreign genetic material, have been bred. These animals help us to understnad what effect a genetic change has on the development in a whole animal. Human genetic disorders certainly will be increasingly addressed by gene replacement therapy; clinical trials for some disorders already have been approved. Genetic recombination in nature is common and essential. It occurs in reproduction and increases our genetic diversity. This process requires homology between strands of DNA and proceeds through a chi (χ) intermediate to generate a DNA with a different sequence.

Chapter 6 Nucleic Acids: How Structure Conveys Information
A Key Figure Revisited

Figure 6.23 The methodology for producing recombinant DNA.

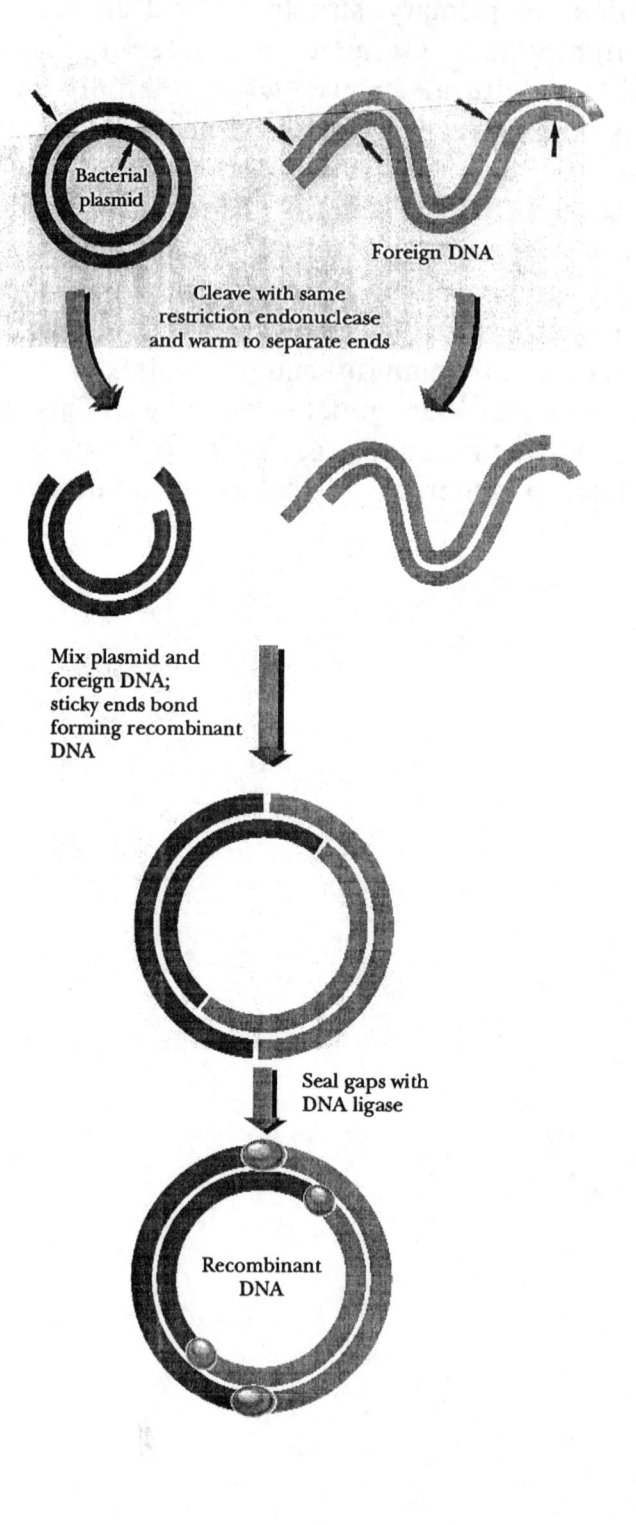

Chapter 6 Nucleic Acids: How Structure Conveys Information

CHAPTER OBJECTIVES

1. Draw the purine and pyrimidine bases, the nucleosides and the nucleotides found in DNA and RNA.
2. Write the numbering scheme for the purine and pyrimidine bases and the sugar unit and draw the phosphodiester backbone that links the nucleotides in nucleic acids.
3. Make a list of the major structural features of DNA and RNA.
4. Describe the hydrogen-bonding interactions in the (GC) and (AT) complementary base pairs.
5. Characterize supercoiled DNA and describe the role of topoisomerases in changing the tertiary structure of circular DNAs.
6. Identify the different types of RNA molecules and characterize each type in terms of their size, abundance and function in the cell.
7. Describe a palindromic recognition sequence and the cutting characteristics of restriction endonucleases and indicate the role of these enzymes in DNA sequencing and recombinant DNA technology.
8. Summarize the relationship between the denaturation of DNA, the melting temperature (T_m) value for the DNA and the observed hyperchromicity.
9. Outline the mechanism for natural genetic recombination.

EXERCISES

1. a. Draw the structure for cytosine. What are the functional groups and the potential reactive sites in this pyrimidine base?
   b. What are the (smaller molecular) components that make up a (ribo)nucleoside?
   c. Name the nucleoside that has an adenine base.
   d. What are the numbers for the positions in the bases and the sugar which are linked in (i) thymidine and (ii) guanosine and what is this linkage called?
   e. What is the number for the ring atom in guanine which has an oxygen bonded to it to form a ketone group?

Chapter 6 Nucleic Acids: How Structure Conveys Information

  f. Name the nucleoside derivative that has the abbreviation AZT. What is the parent nucleoside of this modified nucleoside? What is the modification introduced and at what position does it occur in the nucleoside?

2. There are a variety of nucleotides that have different compositions and structures. Draw the structures for the following nucleotides:
  a. 5'-GDP
  b. 3'-dCMP
  c. 3', 5'-cAMP (cyclic AMP)

3. a. Why is the backbone in DNA or RNA referred to as a phosphodiester backbone?
  b. Describe the structure for the tetranucleotide, pdCpdCpdGpdG, indicating the 5' and the 3' ends.

4. a. The original proposal for the structure of DNA was reported in 1953 by Watson and Crick. This structure was based on a collection of experimental data from many laboratories. One of the most important findings was the fiber diffraction data of Franklin and Wilkins. It was not possible from fiber diffraction data to determine the positions of individual atoms; this information simply had to be implied by effective model building; however, the data revealed a number of distinct distances in the DNA molecule. The most important were 2 nm, 0.34 nm and 3.4 nm. What do these distances correspond to in the DNA molecule?
  b. Perplexing problems faced Watson and Crick as they attempted to build a model for the DNA molecule which was consistent with the data. The bases had to be positioned in a manner that was consistent with their hydrophobic character and arranged in a way that made both chemical and biological sense, together with being consistent with a diameter of 2 nm. Watson and Crick originally assumed that the G, C and T bases existed in the enol (-ene + -ol) tautomeric forms and not the keto forms. Draw the structures for the keto and enol tautomers for guanine.
  c. Draw the (guanosine-cytidine) base pair showing the hydrogen bonds between the bases.

Chapter 6 Nucleic Acids: How Structure Conveys Information

   d. Examine both the (adenosine-thymidine) and the (guanosine-cytidine) base pairs. For the purine base in each base pair, are the hydrogen bonds associated with atoms in the 5- or the 6-membered ring or both ring systems?

   e. If the guanine base were in the enol tautomeric form, would the Watson-Crick hydrogen bonding scheme in the GC base pair be possible?

5. In the early 1980s, Wang and Rich reported the results of a single **crystal** X-ray diffraction study on the synthetic oligonucleotide, d(CGCGCG). This study provided for the first time, crystallographic data which specified the **atomic positions** in the DNA molecule. Data were finally available to support, modify or refute the long held structural proposal of Watson and Crick. The findings were both surprising and stunning! The structure confirmed the hydrogen-bonding scheme that had been proposed by Watson and Crick for a (GC) base pair. However, the DNA exhibited a **left-handed helical twist**. This was the complete opposite to the twist proposed from the fiber diffraction patterns by Watson and Crick. This structural form of DNA was called **Z DNA**. In a very short time following this work, additional X-ray crystallographic studies on other DNAs verified the more common **B form** of DNA, which indeed did exhibit a **right-handed helical twist**.

   Make a list of the structural characteristics of the Z form of DNA and compare these characteristics with those for the A and B forms of DNA.

6. Make a comparative list of differences between DNA and RNA.

7. What is a topoisomerase? What is the function of a topoisomerase in a cell?

8. A viral DNA isolated from monkeys contains 5,000 base pairs.
   a. If the DNA is in the B form, how many base pairs are there in each turn of the helical DNA?
   b. How many complete turns are expected in this DNA?
   c. If this DNA were ligated to make a covalently, closed, circular DNA, it is referred to as the "relaxed" form of

Chapter 6 Nucleic Acids: How Structure Conveys Information

DNA. Explain briefly why the term, "relaxed," is used for this form of DNA.

d. Consider that a topoisomerase cuts this relaxed form of DNA, 10 turns of the helix are removed (the DNA is unwound) and then the DNA is resealed.
  i. Is this a negatively or positively supercoiled DNA?
  ii. Is this a higher or lower energy form of DNA compared to the relaxed form? Explain briefly.

9. The histone proteins that are found in eukaryotic chromatin are some of the most highly conserved proteins known. This means that the amino acid sequence is virtually the same when these proteins are compared in species which diverged at very different times during evolution. Can you suggest a reason why these proteins remained virtually unchanged during millions of years of evolution?

10. There are two billion base pairs in human DNA. If it is assumed that (1) the nucleosome is associated with 200 base pairs of DNA and (2) that all the DNA in the nucleus of the cell is associated with nucleosomes, calculate the number of these repeating nucleosomes units which would be found in human chromatin.

11. a. Name the three different classes of RNA molecules.
    b. What is the function of each class of RNA?
    c. Which class of RNA molecule would be expected to be the most heterogeneous in size? Explain your reasoning.

12. a. What is the difference between the specificity of an exonuclease and an endonuclease?
    b. There have been over 200 restriction endonucleases isolated and characterized. What property is common to all of these enzymes?
    c. A specific restriction endonuclease, *Hae* III, has the recognition sequence shown below.
    GGCC
    CCGG

The hydrolysis of the phosphodiester backbone occurs between the G and the C nucleotides in the middle of the recognition sequence.

Chapter 6 Nucleic Acids: How Structure Conveys Information

How many smaller fragments of DNA would be produced if this restriction enzyme were added to a solution of each of the following DNA molecules? In each case, only one of the DNA strands is shown.
   i.   GCGCATGGCATAGGGCCGGGAAATA
   ii. ATATATGGGGGCATATCCCCGAGAG
   iii. GCGCGGCCGGCCATGGCCATGCGGC

13. DNA can be denatured by increasing the temperature of the DNA solution.
   a. What is the definition of T_m in a DNA denaturation experiment? This experiment is often referred to as a DNA "melting profile."
   b. The melting profile for the same DNA was obtained under two different solution conditions. In one case, the NaCl concentration was 0.001 M, while the other condition was at 0.01 M NaCl. What effect would the different NaCl concentrations have on the T_m value?
   c. The alternating copolymer, poly d(GC) · poly d(GC), can be converted from the B form of DNA to the Z form by a change in the salt concentration in the solution. It is known that the diameter of Z DNA is 1.8 nm, while that for the B form is 2.0 nm. Would you predict the Z form of DNA to exhibit a greater stability at a high or low salt concentration? Briefly explain your reasoning.

14. Indicate whether the following statements are true or false. If the statement is false, correct it.
   a. A (GC) base pair contains three hydrogen bonds, with guanine acting as a hydrogen donor in two of them and a hydrogen bond acceptor in the third hydrogen bond.
   b. The core histones in a nucleosome contain two copies each of H2A, H2B, H3 and H4 histones.
   c. The ultraviolet peak absorbance at 280 nm increases as DNA is denatured.
   d. The class of RNA, which is both the smallest and most homogeneous in size, is tRNAs.
   e. DNA may be spliced by restriction endonucleases and ligated by DNA ligases.
   f. One of the products of DNA recombination, referred to as the single-strand heterozygous product, contains two DNA

Chapter 6 Nucleic Acids: How Structure Conveys Information

molecules. Both of these DNA molecules have one strand, which is the same as the original, while the other strand contains segments of both original DNA strands.

ANSWERS TO EXERCISES

1. a. The positions at which chemical reactions can occur in cytosine are the amine nitrogen atoms (1, 3 and the exocyclic amine), the ketone group at position 2, and the double bond between carbon atoms 5 and 6.

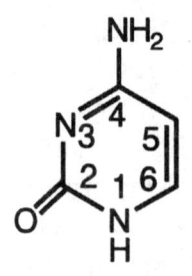

    Cytosine (C)

    b. A nucleoside is made up of a purine or pyrimidine base and a 5-membered cyclic sugar. The sugar in a ribonucleoside is a ribose.

    c. Adenosine

    d. In all nucleosides, the heterocyclic base is linked to the sugar at the C1' carbon atom. In a nucleoside containing a pyrimidine base (C, T or U), the N-1 atom in the base is bonded to the sugar. In the purine nucleosides (A or G), the N-9 atom is bonded to the sugar. In both cases, this carbon-nitrogen bond is called the β-glycosidic bond or linkage.

    e. The ketone group in guanine is at the C-6 position. Note that although the numbering system for both the purine and pyrimidine bases starts at a ring nitrogen, the nitrogen atom at which the numbering begins is different for the purine and pyrimidine base.

    f. AZT is the abbreviation for 3'-azido-3'-deoxythymidine. It is a modified thymidine nucleoside. An azide (N_3^-) group is linked to the C3' position of the deoxyribose sugar.

Chapter 6 Nucleic Acids: How Structure Conveys Information

2a.

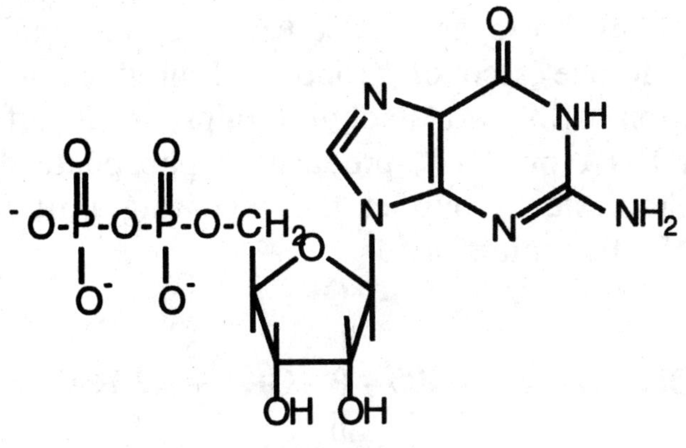

5'-GMP

b.

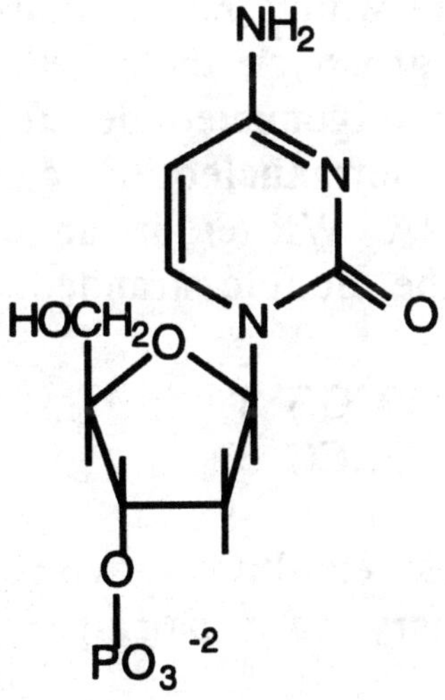

3'-dCMP

c.

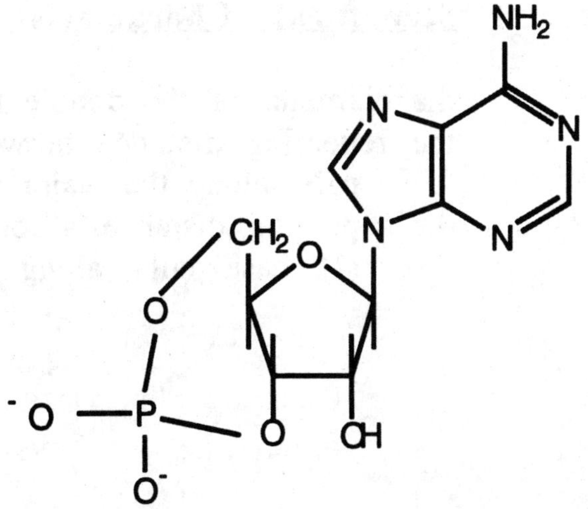

3', 5'-cAMP

Chapter 6 Nucleic Acids: How Structure Conveys Information

3. a. The reaction of an acid with an alcohol produces an ester and water. Phosphoric acid is a triprotic, inorganic acid. Reaction with one alcohol group will produce a phosphate ester. Reaction with two alcohol groups, as is effectively the case in DNA or RNA, produces a phosphate diester. This phosphodiester group is the repeating unit in the backbone of all nucleic acids.

$$H_3PO_4 + 2\,R\text{-}OH \longrightarrow RO\text{-}\underset{\underset{O}{\|}}{\overset{\overset{O^-}{|}}{P}}\text{-}OR + 2\,H_2O$$

   b. The tetranucleotide, 5'-pdCpdCpdGpdG-3', is a small **deoxy**ribonucleotide, which has a phosphate group on the 5' end and an alcohol group on the 3' end. It is also an example of a self-associating oligonucleotide. If dissolved in solution, this single-stranded tetranucleotide can spontaneously form a double-stranded DNA. Written in an even more abbreviated form, the double-stranded DNA would be:

   5' CCGG 3'
   3' GGCC 5'

   Note: This molecule exhibits a two-fold symmetry, which permits complementary base pairing in the formation of the double helix.

4. a.

Distance	Structural Characteristic
2 nm	the diameter of the double helical DNA
0.34 nm	the repeating distance between the base pairs along the major axis of DNA
3.4 nm	the repeating distance of one complete turn (10 base pairs) along the length of the DNA

Chapter 6 Nucleic Acids: How Structure Conveys Information

b.

Keto-enol equilibrium for guanine, in which the keto tautomer predominates.

c.

Complementary base pairing between guanosine and cytosine (R= ribose).

Chapter 6 Nucleic Acids: How Structure Conveys Information

d. In the Watson and Crick base-pairing scheme, the hydrogen bonds to or from the guanine (GC base pair) or adenine base (AT or AU base pair) are associated with the atoms in (only) the **six-membered** ring.

e. With the guanine in the enol form, the Watson-Crick base pairing scheme can not occur.

5.

Characteristic	Z DNA	B DNA	A DNA
Helical twist	left handed	right handed	right handed
Separation between bases	(0.37 nm)*	0.34 nm	(0.26 nm)
Anti-parallel strands	yes	yes	yes
Tilt to base pairs	slight	very slight	20^0 tilt
Number of base pairs/turn	(12)	10	11
Grooves on outside of DNA	barely detectable	very distinct	distinct
Diameter	(1.8 nm)	2.0 nm	(2.6 nm)
Uses Watson-Crick base pairing	yes	yes	yes

* The values in parentheses are not presented in the text.

Chapter 6 Nucleic Acids: How Structure Conveys Information

6.

	DNA	**RNA**
Strandedness	double	single
Functional participation	replication and transcription	transcription and translation
Composition		
Bases	A, T, G, C	A, U, G, C
Ribose	2'-deoxyribose	ribose
Location in cell	nucleus	nucleus, cytoplasm and within ribosomes

7. A topoisomerase is a cellular enzyme that acts on a circular DNA substrate to change its level of supercoiling. In this reaction the DNA backbone is cut, an additional twist is put in or taken out, and then the backbone is resealed. When this action occurs, the resultant DNA is in a different topology and one topological isomer of DNA has been converted to another one (DNA with a different level of supercoiling). This explains the reason for the name and classification, topoisomerase, for the enzyme.

    The function of a topoisomerase in the cell is to maintain the level of superhelicity in the DNA. It has been shown in a number of cases that replication, transcription and recombination processes are sensitive to the superhelicity of the DNA.

8. a. 10 base pairs per turn
   b. [5000 base pairs]/[10 base pairs per turn] = 500 helical turns
   c. There is no additional stress or torsional forces on the DNA as a result of making the molecule circular.
   d. i. Negatively supercoiled DNA contains fewer turns than that found in the B form of DNA (490 turns < 500 turns). This is a negatively supercoiled DNA.
      ii. This is a higher energy form because the removal of helical turns puts an additional stress on the DNA. To compensate for the stress, the DNA changes its tertiary structure to form additional turns in the form of supercoils.

Chapter 6 Nucleic Acids: How Structure Conveys Information

9. The conservation or lack of change in the amino acid sequence in specific proteins throughout evolution is thought to be an indication of the essential role of the protein in processes vital to life. The histones are among the most highly conserved proteins known. They interact directly with DNA and form nucleosomes, which are the fundamental repeating subunits involved in the packaging of DNA and in the formation of chromatin structure in the nucleus. Any changes in the amino acid sequence must have an extremely detrimental effect on the packaging of DNA and therefore on the life of the cell.

10. Within the assumptions in the question, the number of nucleosomes in the nucleus of a human cell is:

    [$2.0 \times 10^{+9}$ base pairs]/[$2.0 \times 10^{+2}$ base pairs per nucleosome] =

    $1.0 \times 10^{+7}$ nucleosomes in human chromatin

11. a. There are three classes of RNA molecules:
    i. tRNA- transfer RNA
    ii. rRNA- ribosomal RNA
    iii. mRNA- messenger RNA

    b. Each tRNA molecule complexes with a specific amino acid at the 3' end of the tRNA and activates that amino acid for the translational process on the ribosomes.
    Ribosomes are made up of rRNA molecules and a large number of proteins. These ribosomes provide the "stage" or "the work bench" where protein synthesis takes place.
    The mRNA molecules are the molecular message carriers, which take the genetic message to the ribosomes. The sequence of the mRNA therefore determines the composition and amino acid sequence of the protein to be synthesized.

    c. The most heterogeneous class of RNA is mRNA. There are a multitude of genes of widely varying sizes that are transcribed. The size of the gene will generally reflect the size of the protein to be synthesized.

12. a. An exonuclease digests nucleic acids by the progressive hydrolysis of the terminal nucleotide. An endonuclease digests

Chapter 6 Nucleic Acids: How Structure Conveys Information

nucleic acids by hydrolyzing the backbone in the interior of the molecule.

b. Each restriction endonuclease recognizes a specific sequence of double-stranded DNA, binds to it, and catalyzes the hydrolysis of the phosphodiester backbone at specific cut sites. The recognition sequence exhibits a two-fold symmetry and contains four or more base pairs.

c.
i. There is only one GGCC recognition site in this molecule; therefore, there is only one site for binding and subsequent hydrolysis. Two DNA fragments are produced, one with 10 base pairs and the other with 15 base pairs.

ii. There are no GGCC recognition sites for this restriction endonuclease in this DNA; therefore, there is no enzymatic cleavage and no smaller DNA fragments are produced.

iii. Four DNA fragments are produced because there are three GGCC recognition sites. The DNA fragment sizes produced have 4, 6 and 10 base pairs. Note that two different DNA fragments with 6 base pairs are produced.

13.
a. The T_m value is the temperature at which the DNA is partially denatured and the ultraviolet absorbance at 260 nm is halfway between the final and starting absorbance values.

b. Increasing the salt concentration increases the stability of DNA and therefore also increases the value of the melting temperature. There are a number of forces that determine the stability of DNA. A force that tends to destabilize double stranded DNA is the repulsive force between the two DNA strands as a result of the negative phosphate units on each strand. The greater stability of DNA at a higher salt concentration results because the positively charged Na^+ ions bind to the negatively charged phosphates on the DNA backbone and reduce the effective electrostatic repulsion of the interstrand phosphates.

c. The Z form of DNA has a smaller diameter. The interstrand phosphate units are closer to each other in Z DNA than in the B DNA. At low salt concentration, the B DNA will be more stable because of the smaller electrostatic repulsive forces between strands. The Z form is expected to be more stable at a high salt concentration since this condition would favor a reduced electrostatic repulsion between the interstrand phosphates.

Chapter 6 Nucleic Acids: How Structure Conveys Information

14.
- a. True.
- b. True.
- c. False. The ultraviolet absorbance peak for DNA occurs at 260 nm. The intensity of this peak does increase on converting native DNA into denatured DNA.
- d. True.
- e. False. Splicing means to join together. Restriction endonucleases catalyze the cleavage or cutting of double-stranded DNA, while DNA ligase catalyzes the splicing of (covalently links) DNA strands together.
- f. True.

7

Nucleic Acid Biotechnology Techniques

The ability to manipulate DNA has created an explosion of opportunities in fundamental biochemical and biomedical research and in the potential for direct clinical applications to many human diseases. The discovery of restriction endonucleases provides the basis for isolating specific DNA fragments or genes from an organism. The DNA fragments can be separated using **polyacrylamide gel electrophoresis** (PAGE). Very small amounts of the fragments can be detected by labeling the DNA with radioactive phosphorus (^{32}P), using **autoradiography** detection. By using these fundamental findings, the primary sequence of a gene or DNA segment can be determined. The **Sanger-Coulson** method for sequencing DNA is a widely used procedure based on the selective interruption of oligonucleotide synthesis by incorporation of dideoyribonucleoside triphosphates. A DNA segment from another organism can be incorporated into a **bacterial plasmid** or bacteriophage to form a **chimeric DNA** and this recombinant DNA **cloned**. The entire genome of an organism can be cloned to produce a **DNA** (or genomic) **library** for the organism or alternatively, a cDNA library can be generated, being derived from the

Chapter 7 Nucleic Acid Biotechnology Techniques

mRNA expressed in a specific organ. The **polymerase chain reaction** (PCR) is a recent breakthrough of enormous importance in which the amount of a specific DNA fragment is greatly increased without the need for cloning. The location of genes within the genome of an organism are now being determined using both classical genetics and physical mapping procedures. **Restriction-fragment-length polymorphism** (RFLP) provides a mechanism for finding the location genes within the genome and is becoming a valuable tool in forensic science. Recombinant DNA can be manipulated so that bacteria can serve as "cellular factories" for producing large amounts of recombinant proteins, with the production of human recombinant insulin being a well known example. Transgenic animals have been produced in which a gene from another organism is incorporated into the germ-line of the animal. Currently there are efforts in human gene therapy in which a normal gene is introduced into somatic cells to correct a genetic disease caused by a missing protein.

Chapter 7 Nucleic Acid Biotechnology Techniques

Some Key Figures Revisited

Figure 7.7 The cloning of human DNA fragments with a viral vector. Human DNA is inserted into viral DNA and then cloned.

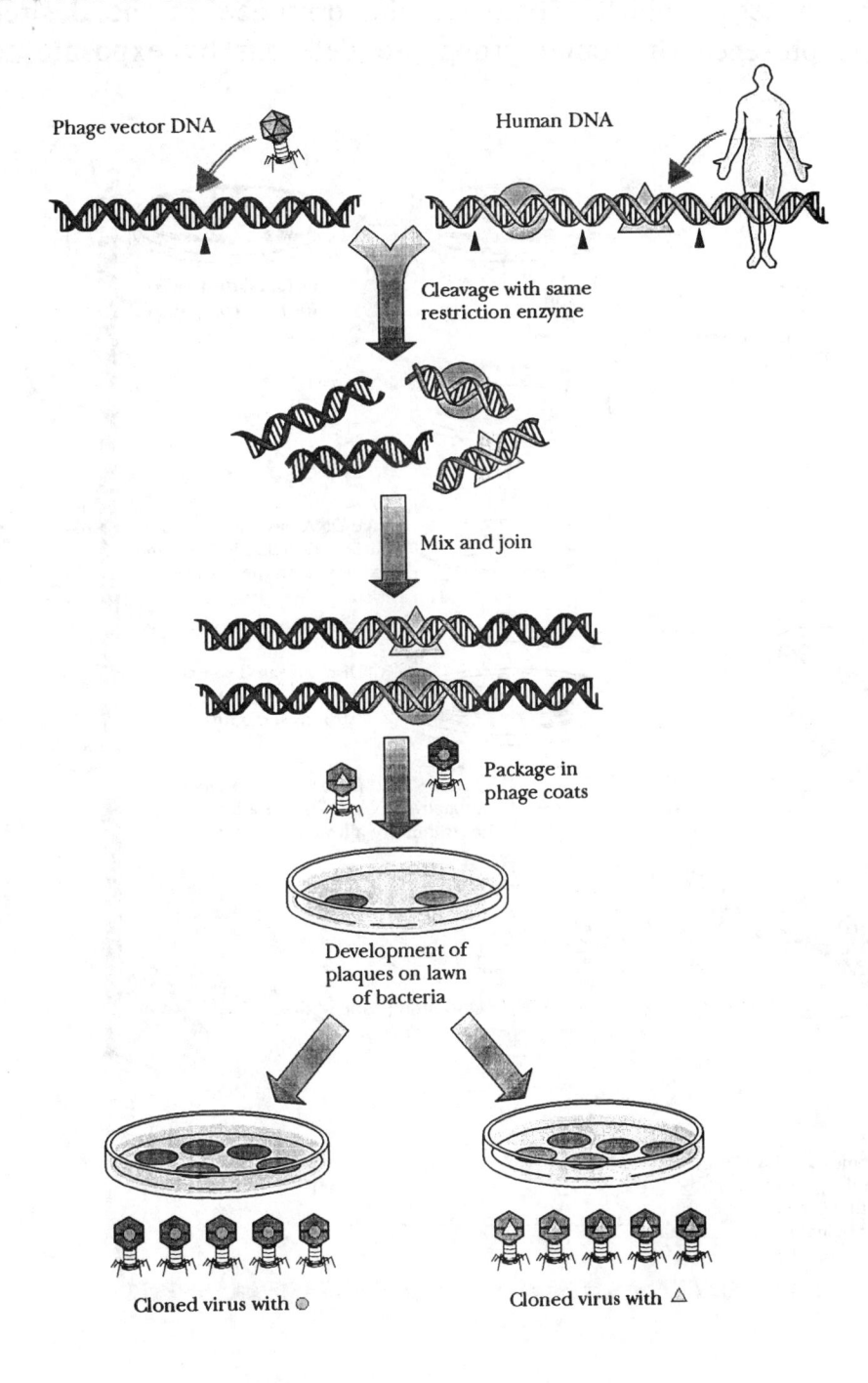

Chapter 7 Nucleic Acid Biotechnology Techniques

Figure 7.13 Selecting a desired clone from a DNA library. A portion of each clone is transferred to a nitrocellulose disc by blotting. The disc is treated with a denaturing agent to unwind all the DNA. A single-stranded radioactive probe for the desired DNA is added and allowed to anneal. After excess probe is removed, the presence of the desired clone and the presence of bound probe are detected by exposure to x-ray film.

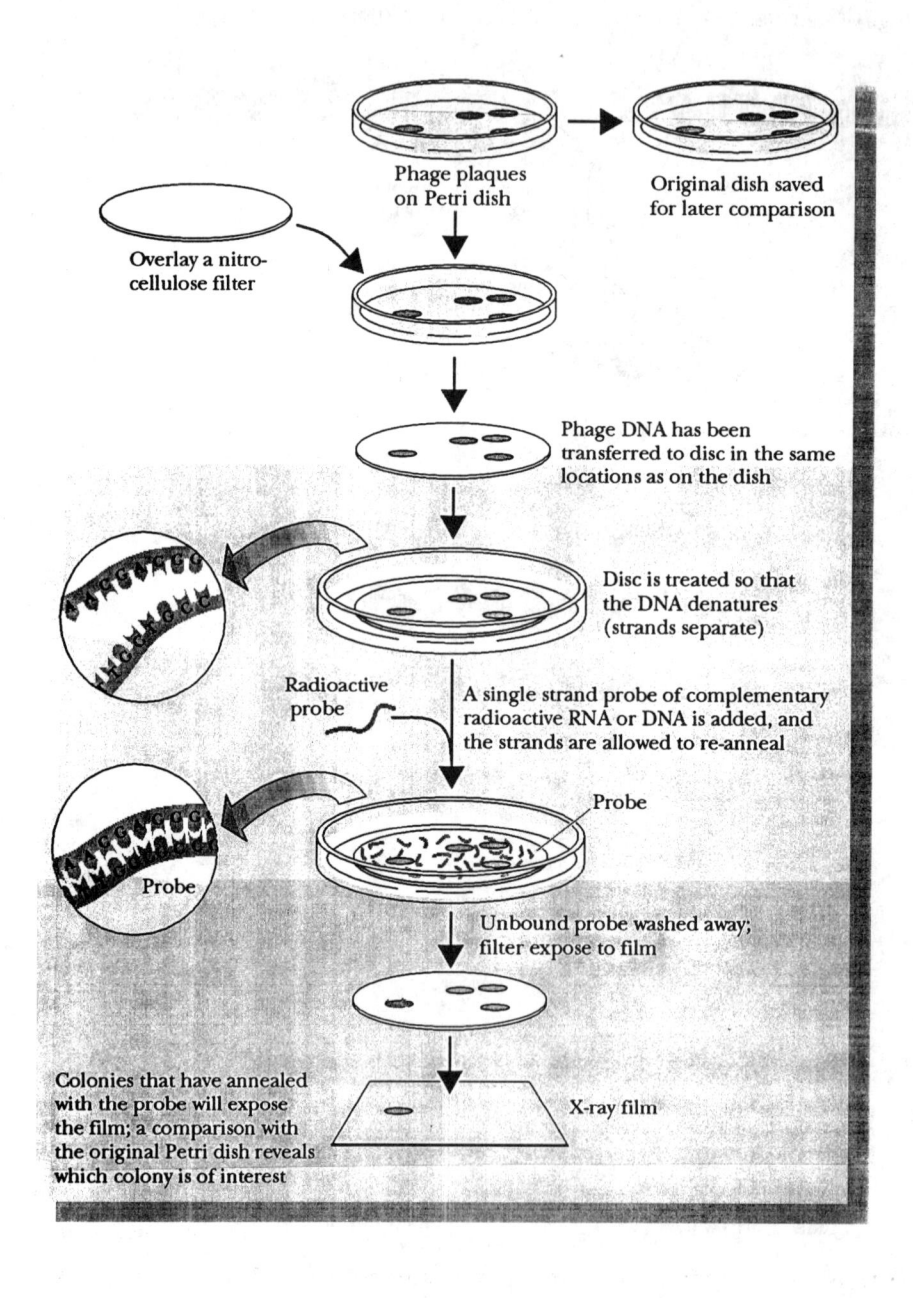

Chapter 7 Nucleic Acid Biotechnology Techniques

CHAPTER OBJECTIVES

1. Indicate the parameters that determine the extent to which DNA fragments can be separated on polyacrylamide gel electrophoresis (PAGE).
2. Outline the procedure by which the primary sequence of DNA can be obtained using the Sanger and Coulson method.
3. Outline the procedure used in cloning a DNA fragment in both a bacterial plasmid and a bacteriophage, pointing out the role or nature of restriction endonucleases, DNA ligase, bacteria lawns, clones and plaques.
4. List the distinguishing characteristics of a cDNA library and a genomic (DNA) library.
5. Describe the procedure used in the polymerase chain reaction (PCR), pointing out the essential roles of primers, *Taq* DNA polymerase and temperature recycling.
6. Indicate how classical genetic mapping and physical mapping are used to construct a molecular map for the genes on a chromosome.
7. Discuss the usefulness of restriction-fragment-length polymorphism (RFLP) in the diagnosis of disease.
8. Outline the role of gene therapy directed toward somatic cells in the treatment of genetic diseases.
9. Describe the procedure used to produce a transgenic animal and indicate how this differs from gene transfers into somatic cells.

SPOTLIGHT

DNA in Court:
"Molecular Fingerprinting" Evidence

A fight erupts in an apartment, a shot is fired, the next-door neighbor dials 911, and an hour later, the police file a homicide report. A suspect is arraigned and pleads innocent. There are no witnesses and only a precious few pieces of evidence appear useful to the prosecution. A strand of hair, the same color as that of the suspect, was found under the fingernail of the victim. A modern forensic scientist takes this microgram specimen, analyzes the DNA in it and determines if

Chapter 7 Nucleic Acid Biotechnology Techniques

it is identical to the DNA isolated from the suspect. If it is, this will provide virtually indisputable evidence that the suspect is the murderer.

The basis for this "molecular fingerprinting" lies in the finding that there are small segments of DNA (a few hundred base pairs) which occur repeatedly, from 4 to ca. 500 times, in humans. These sequences [called "minisatellite DNA" (MS DNA)] are similar in all persons, but are unique for each individual, due to different mutations (changed bases). The DNA can be digested with a restriction enzyme to produce a number of DNA fragments of different sizes. The number and the sizes of these DNA fragments will be dependent on the number of copies of MS DNA and the specific mutations in the DNA of each individual. Consider the DNAs in Figure 1 from three different individuals.

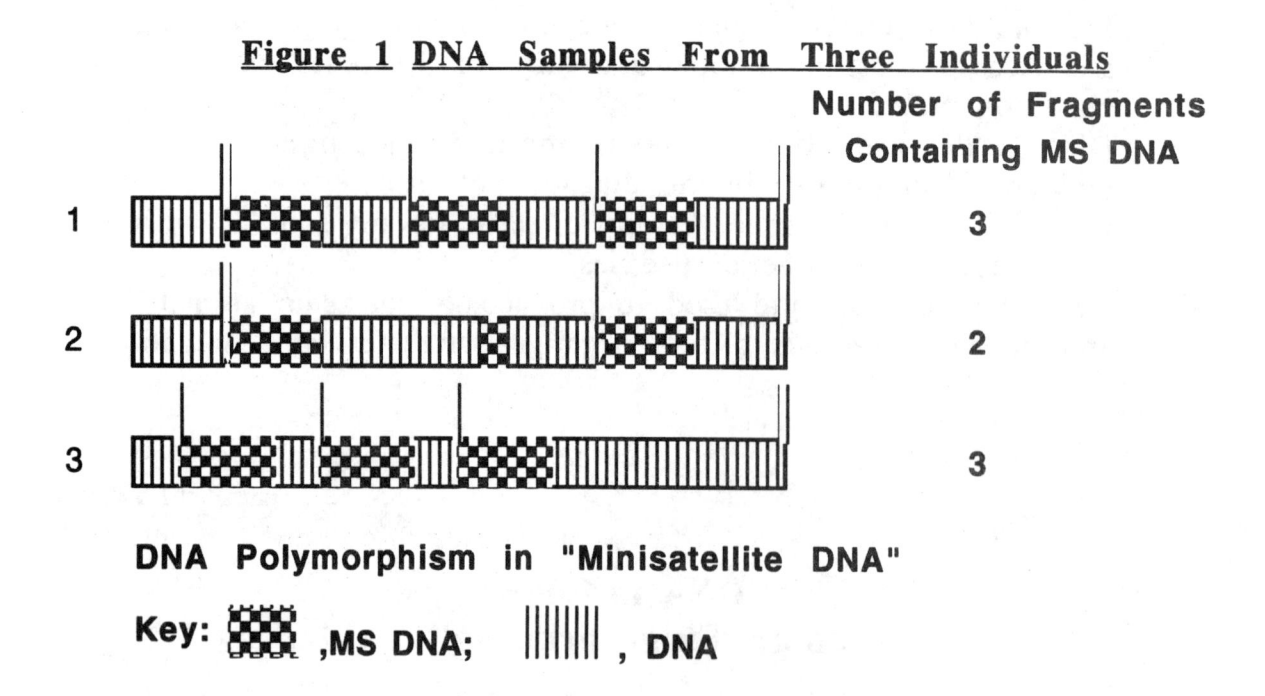

Figure 1 DNA Samples From Three Individuals

Number of Fragments
Containing MS DNA

1 — 3
2 — 2
3 — 3

DNA Polymorphism in "Minisatellite DNA"

Key: ▦ ,MS DNA; ||||||| , DNA

Notice that all three DNAs contain multiple copies of the MS DNA. Sample 2, however, contains a mutation that results in the deletion of the 5'-end of the second MS DNA. Sample 3 has multiple mutations that result in a large deletion of DNA between the MS DNA repeats. A restriction enzyme, such as Eco RI, is used to digest these DNA samples. The <u>recognition sites occurs at the 5'-end of each intact MS DNA segment</u> and

Chapter 7 Nucleic Acid Biotechnology Techniques

at the 3'-end of the DNAs shown. After digestion, the DNA fragments are separated by gel electrophoresis (Figure 2). Because of the inherent uniqueness in the sequence of each DNA, a distinct pattern of bands isobserved for each DNA sample. Although millions of bands is produced in the digestion of human DNA, the interest is on only those fragments that contain the MS DNA.

To focus on and detect just the size and the number of DNA fragments in the sample that contain the MS DNA, a simple, yet powerful procedure is necessary. The steps in this "molecular fingerprinting" procedure are outlined in Figure 2.

First, the gel is treated with NaOH to denature the double-stranded DNA. This procedure disrupts the hydrogen bonding between the complementary strands to produce single-stranded DNA.

After neutralization, the DNA is transferred to a filter paper that binds the single-stranded DNA. The filter retains the DNA fragments in the identical positions as that in the gel, but additionally places the DNA on the surface of the filter paper. Molecular biologists refer to this transfer procedure as a Southern blot.

To detect the specific bands that contain the MS DNA, a DNA fragment, which is complementary to the MS DNA, is radioactively labeled and then denatured to make a so-called single-strand DNA "probe."

The filter paper is then immersed in a solution containing the radioactive probe. The probe will hybridize to only complementary MS DNA sequences, which can then be detected by autoradiography.

Figure 3 shows the final results (autoradiogram) that are observed for the DNA from three individuals (noted in Figure 1), and the DNA isolated directly from the hair found under the fingernail of the victim. DNA Sample 2 was taken from the suspect, while Samples 1 and 3 were from other individuals. You be the judge. Based on the DNA fingerprinting data, is the suspect guilty or innocent?

Chapter 7 Nucleic Acid Biotechnology Techniques

Figure 2 Detection Of MS DNA Using Southern Blotting

A human hair → Isolate DNA → Digest with Eco RI → Separate fragments by gel electrophoresis → Transfer to Filter paper (Southern blot) → Millions of bands appear as a smear → Hybridize with radioactively labeled minisatellite DNA → Autoradiogram

Bands for DNA fragments which hybridize to minisatellite DNA

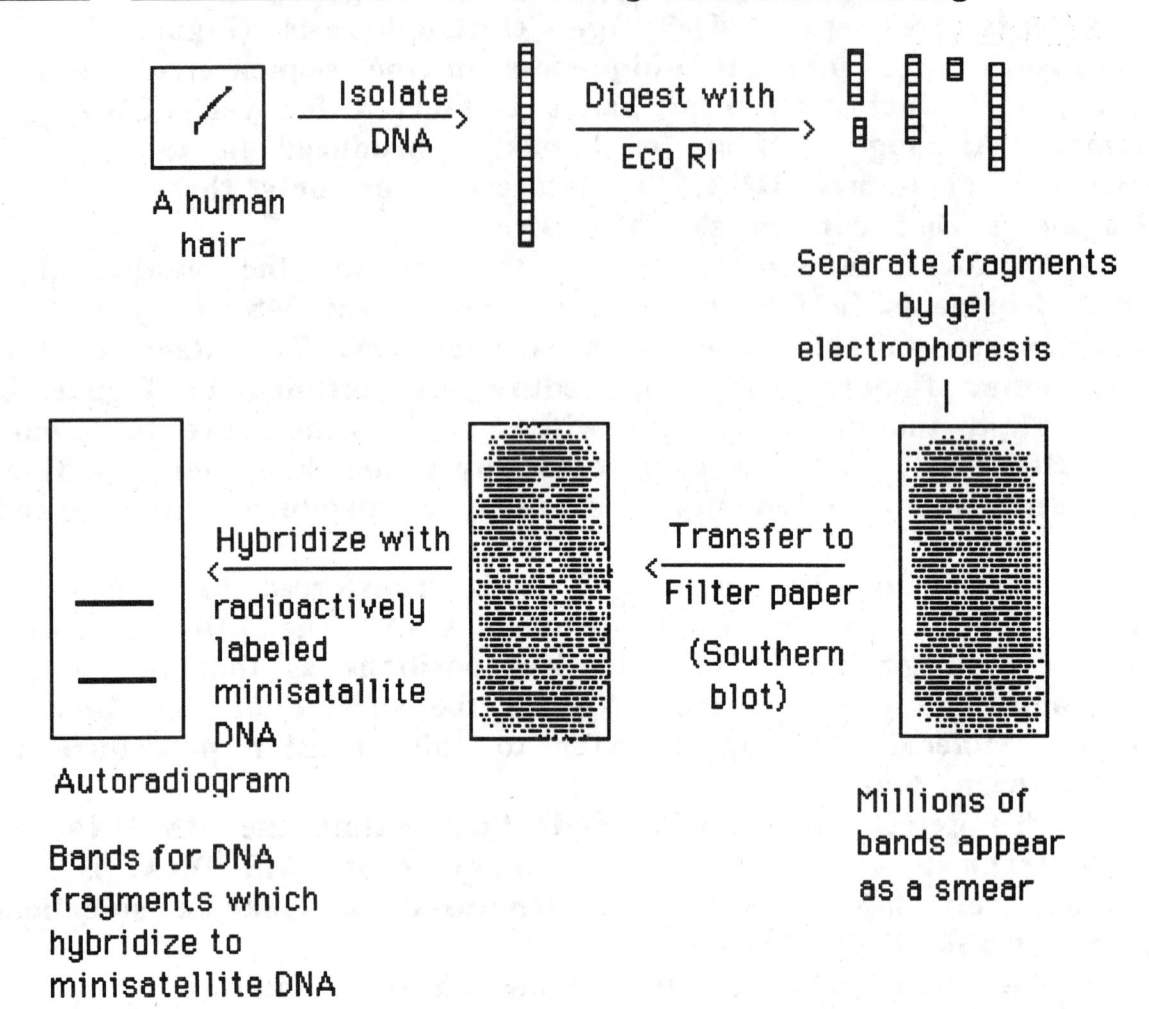

Figure 3 Autoradiography For DNA Samples

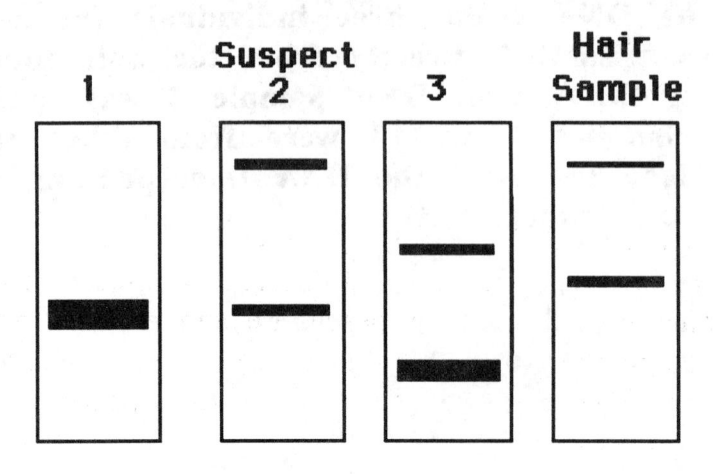

Suspect 1, Suspect 2, Suspect 3, Hair Sample

96

Chapter 7 Nucleic Acid Biotechnology Techniques

EXERCISES

1. A circular, highly supercoiled, simian viral DNA of 5000 base pairs is isolated from an infected cell. A portion of the sample is cleaved with a restriction enzyme, which makes a single cut, to produce a linear DNA. Gel electrophoresis is carried out on the supercoiled and the linear DNA samples. What is observed on the gel?

2. In electrophoresis of proteins, the protein samples are added to SDS (sodium dodecylsulfate) and a reducing agent, such as β-mercaptoethanol, and then boiled for a few minutes before gel electrophoresis is carried out. In the electrophoresis of DNA, DNA samples are loaded onto the gel without a similar preliminary sample preparation. Explain why the procedures are different for proteins and for DNA.

3. Consider an experiment in which a fragment of human β-globin gene is cloned into the DNA of a bacteriophage, which then is used to infect a bacteria cell.
    a. Indicate the vector DNA and the composition of the chimera DNA.
    b. Indicate what is meant by a clone, a lawn of bacteria and a plaque.
    c. How is the fragment of DNA eventually separated from the bacteriophage DNA?

4. How does a cDNA library from a human liver cell differ from a human DNA (genomic) library?

5. If you knew that a protein were expressed in a liver cell, would you initially screen the cDNA library or the DNA (genomic) library from the liver to isolate the gene of interest?

6. An unexperienced biochemist attempts to isolate a gene from a cDNA library that contains chimeric bacterial plasmids. The biochemist uses a blotting procedure and a DNA probe to determine which colony contains the gene of interest. However, a number of mistakes are made in the procedure. Indicate the

Chapter 7 Nucleic Acid Biotechnology Techniques

mistakes and specify how each of these steps should be corrected so that this young scientist accomplishes the goal.

A piece of nitrocellulose filter paper is placed over a lawn of individually separated bacteria in a Petri dish. At least two distinct marks are made on the nitrocellulose and the bacterial lawn so that the location of the clone of interest can be identified after the procedure. After enough time for the bacteria to grow on the nitrocellulose, it is removed for probing and the Petri dish discarded.

The colonies on the nitrocellulose are then lysed and the DNA denatured. This is then probed with a double-stranded DNA that contains the complementary sequence to the gene of interest. The nitrocellulose is then removed after sufficient time for hybridization to occur. The excess solution is removed and then the nitrocellulose filter is placed on the bench-top overnight so that a color change in the colony develops at the correct clone. The next morning, he observes that the nitrocellulose remained the same as the day before and that no spots are detectable.

7. A restriction enzyme, Eco RI, is used to restrict the human genome. Eco RI cuts the DNA about every 4100 bps, on average. Approximately how many DNA fragments are produced in this digestion? Assume that the human genome contains about three billion base pairs.

8. There are DNA sequences found in the cluster of human β-globin genes, which are referred to as pseudogenes. How does a pseudogene differ from the β-globin gene?

9. Bacterial conjugation studies are used to determine the relative location of genes A, B and C in the genome of a donor cell. The findings in the recipient cell indicate that genes B and C are close to each other, while gene A is far from B or C.
    a. Briefly describe the experimental approach used to obtain these findings.

    b. What is the observation which indicates that genes A and B are in close proximity to each other, while gene C is far from the other two on the chromosome?

Chapter 7 Nucleic Acid Biotechnology Techniques

10. A colleague of yours is interested in cloning a gene in a bacterial plasmid, as shown in Figure 7.9 in the text. He successfully inserts the foreign gene in the plasmid, but the insert is directly in the center of the gene for antibiotic resistance for ampicillin. He adds this recombinant DNA to a suspension of bacteria, in which some of the bacteria take up the plasmid and propagate with the foreign gene. He tries to grow a large batch of bacteria in a medium that contains ampicillin, but is unsuccessful. Indicate the mistake that occurred in this experiment.

11. Name at least one major health problem which can, in theory, be corrected with gene therapy and one which can not be corrected.

12. How does the result of gene therapy that has occurred in germ-line cells fundamentally differ from that in somatic cells?

13. Indicate whether the following statements are true or false. If a statement is false, correct it.
    a. A double-stranded DNA can be used as a primer in PCR.
    b. Autoradiography is used to detect the clones of interest in a DNA library.
    c. DNA is separated on polyacrylamide gel electrophoresis by both the size and the magnitude of the charge.
    d. ^{32}P is the most commonly used radioactive isotope in the labelling of DNA.
    e. The Sanger and Coulson procedure for sequencing DNA depends on the incorporation of a modified deoxyribonucleoside triphosphate, which has a hydrogen in the 2' position, into the synthesized DNA. This incorporation terminates the synthesis of that strand.
    f. After 20 cycles of PCR on a DNA fragment, the amount of DNA has increased by about 10,000.
    g. Gene clusters can be mapped by the use of chromosomal walking.
    h. A plasmid carries two genes for drug resistance, the ampicillin and the tetracycline resistance genes. If a rat gene to be cloned is inserted directly into the tetracycline resistance gene, the resulting bacterium cell which contains

Chapter 7 Nucleic Acid Biotechnology Techniques

this chimeric DNA will exhibit tetracycline resistance, but not ampicillin resistance.

ANSWERS TO EXERCISES

1. For linear DNA fragments, polyacrylamide gel electrophoresis separates DNA by size, with smaller DNA fragments exhibiting the greater mobility. This separation occurs because the gel is effectively a matrix of "tunnels," which the DNA "snakes" through, migrating from the negatively charged end of the gel toward the positively charged end. In the case of supercoiled DNA, the DNA is circular and coiled-up. This change in the shape alters its mobility characteristics. Since the supercoiled DNA is more compact and smaller than linear DNA, it has an increased mobility in the gel relative to the linear DNA.

2. Proteins migrate in SDS-PAGE in reducing conditions (in the presence of β–mercaptoethanol) according to their molecular weight, with the mobility being inversely related to the molecular weight. The β–mercaptoethanol is required to ensure that all the cysteines are in the reduced form and are not involved in intrastrand cross-links. The SDS, an ionic detergent, and the boiling of the sample, insures that the protein is completely denatured to form an extended random coil.

   There is no need for β–mercaptoethanol in a double stranded DNA since the redox state is not in question or a concern. No SDS is added and the samples are not heated because the DNA must remain in the double-stranded form. These procedures encourage denaturation of the DNA and produce denatured single-stranded DNA.

3. a. The vector is the bacteriophage DNA, with the chimeric DNA containing the human β–globin gene ligated into the bacteriophage DNA.
   b. A clone is a genetically identical population of DNA, cells or organisms; a lawn of bacteria refers to the collection of

Chapter 7 Nucleic Acid Biotechnology Techniques

bacterial colonies on a Petri plate; the bacteriophage infects a cell and as a result, the cell dies and more bacteriophage are produced. An individual colony of bacteria (a clone) contains millions of bacteria cells originating from the same original cell. As the infection progresses to many cells in the colony, a plaque is observed and marks the location of the killed bacteria.

4. A cDNA originates from the population of mRNAs in a specific cell type; therefore, the cDNA library from a liver cell contains a very limited number of genes, with only those expressed in the liver being present. A cDNA library from a different organ will also be composed of a limited number of genes and will be quite different than that of the liver.

   On the other hand, a DNA (genomic) library contains all the DNA in the cell; therefore, a genomic library is much larger than a cDNA library. It contains all the genes in the organism and all other DNA, much of which has an unknown function at this point in time. Since a genomic library contains all the DNA in the genome of the organism, a DNA library can be prepared equally well from any cell in a multicellular organism.

5. If it is known that a protein is expressed in the liver cell, initial screening should always be done on the cDNA library since it is so much smaller (contains fewer clones) and the screening of the clones can be accomplished in a shorter time.

6. Mistakes made and corrections.
   a. The Petri dish containing the bacteria should not be discarded since this contains the clones. Since the colonies on the nitrocellulose are used to screen the colonies, they will be destroyed in the process.

   b. The nitrocellulose must be probed with single-stranded DNA. Double-stranded DNA will not hybridize to the complementary sequences on the nitrocellulose. Also, the single-stranded DNA used as a probe must be radiolabeled with ^{32}P to provide the detection system for the screening. In practice, the double-stranded DNA is labelled with the ^{32}P and then the labelled DNA is denatured to produce the single strands.

Chapter 7 Nucleic Acid Biotechnology Techniques

    c.   There will be no color change since there was no method for detection utilized. The detection is carried out by

autoradiography. The nitrocellulose which was probed with the radiolabeled DNA is rinsed and dried and then placed with a photographic film. The position of the clone of interest will be evident in a spot being produced on the photographic film at the specific location. This position on the nitrocellulose filter directly corresponds to a colony that contains the gene of interest (in the Petri plate).

7.   Assuming a more or less random cutting pattern, Eco RI cuts DNA about every 4100 bps. In the human genome of about three billion bps, the number of fragments produced is about

$$\frac{3 \times 10^{+9}}{4.1 \times 10^{+3}} = \text{approximately } 7 \times 10^{+5} \text{ DNA fragments}$$

8.   A pseudogene is not expressed at any time in the life of the organism. It contains a mutation, which effectively inactivates the gene. The β-globin gene is expressed during the development of the organism.

9.   a.   The donor and recipient bacteria are mixed in suspension and during physical contact DNA is slowly transferred from the donor to the recipient cells. This procedure is carried out for various times to permit different size segments of DNA to be transferred. The recipient cells are separated and the DNA that is transferred from the donor cells into the recipient cells is analyzed.

    b.   In the simplest case, when the transferred DNA is analyzed, genes A and B are found together indicating that they are ransferred in the same time period. Gene C is transferred in cases in which no A and B are found or alternatively, it is found with A and B only after much longer contact times. This indicates that genes A and B are in close proximity in the DNA, while gene C is located at a considerable distance from genes A and B.

Chapter 7 Nucleic Acid Biotechnology Techniques

10. The investigator cloned the foreign gene into the gene for antibiotic resistance. This procedure completely inactivates the ampicillin gene; therefore, cells that have incorporated the recombinant and cells that did not take up the recombinant will be ampicillin sensitive (they will not exhibit antibiotic resistance). All the cells die in the medium. The procedure of inserting a DNA fragment into a functioning gene, which results in the elimination of its activity, is referred to as INSERTIONAL INACTIVATION. This procedure is often useful when the plasmid contains two different antibiotic resistance genes, such as one for ampicillin and tetracycline. In this case, the cloning inactivates the resistance to one of the antibiotics, while the resistance to the other antibiotic is retained.

11. Health problems that can be corrected by gene therapy are genetic in origin, such as cystic fibrosis, SCID, muscular dystrophy and some forms of cancer. Health problems, such as bacterial infections, cannot be corrected by gene therapy.

12. The important difference is that gene therapy carried out in a germ-line cell incorporates the gene of interest in all cells in the developing organism. This gene is then be passed onto all future generations. On the other hand, gene therapy in a somatic cell can correct the health problem for the organism, but it is not passed onto future generations.

13.
    a. False. DNA primers used in PCR must be single stranded since the primers must hybridize with the target DNA. Double-stranded DNA must be completely dissociated into individual single strands before use.
    b. True.
    c. False. The DNA is separated by size only.
    d. True.
    e. False. The modified nucleotide will have a hydrogen in the 3' carbon of the deoxyribose group instead of a hydroxyl group.
    f. True.
    g. True.
    h. False. The cell will exhibit ampicillin resistance and no tetracycline resistance. The tetracycline resistance has been lost by insertional inactivation of the rat gene within the tetracycline resistance gene.

8

Lipids and Membranes

    Lipids are a diverse family of biomolecules, collectively classified as such because of their solubility in non-polar solvents. Some of the major lipid molecules are fatty acids and the carboxylic- and phosphatidyl- esters of glycerol and sphingosine. The variety and complexity of the lipids can be increased by reaction of a phosphatidic acid with the alcohols, ethanolamine, choline, serine or inositol. These resultant amphiphilic molecules are the major constituents of biological membranes. The ester linkages can be hydrolyzed chemically or alternatively by cellular enzymes called lipases. **Steroids** represent another major class of lipids, which includes cholesterol, male and female sex hormones, vitamins and bile salts. The fat soluble vitamins, including vitamins A, D, E, and K, serve in a variety of essential biological functions. Two groups of powerful physiological agents, the **prostaglandins** and **leukotrienes**, both containing 20 carbon atoms, are derived from arachidonic acid and are also members of the lipid family of molecules.
    Biological **membranes** serve as actively functioning envelopes around both the cell and the organelles within them. These molecular assemblies provide both the means for communicating with other cells in unicellular and multicellular organisms (or between organelles in a

Chapter 8 Lipids and Membranes

eukaryotic cell) and the mechanisms by which ions, nutrients, waste products and even macromolecules enter and leave these metabolizing units. The basic components of all membranes are simply **lipid bilayers**; however, a mosaic of structural and functional variations exists between organelles and between cells from different tissues. These variations are due to the presence of different lipid molecules, functional protein receptors, transport proteins and gated-channels associated with the bilayer, all of which contribute to making the molecular assembly a functioning biological membrane for a living cell.

A Key Figure Revisited

Figure 8.21 (next page) The mode of action of the LDL receptor. A portion of the membrane, with LDL receptor and bound LDL, is taken into the cell as a vesicle. The receptor protein releases LDL and is returned to the cell surface when the vesicle fuses with the membrane. LDL releases cholesterol in the cell. An oversupply of cholesterol inhibits synthesis of the LDL receptor protein. An insufficient number of receptors leads to elevated levels of LDL and cholesterol in the bloodstream. This situation increases the risk of heart attack.

Chapter 8 Lipids and Membranes

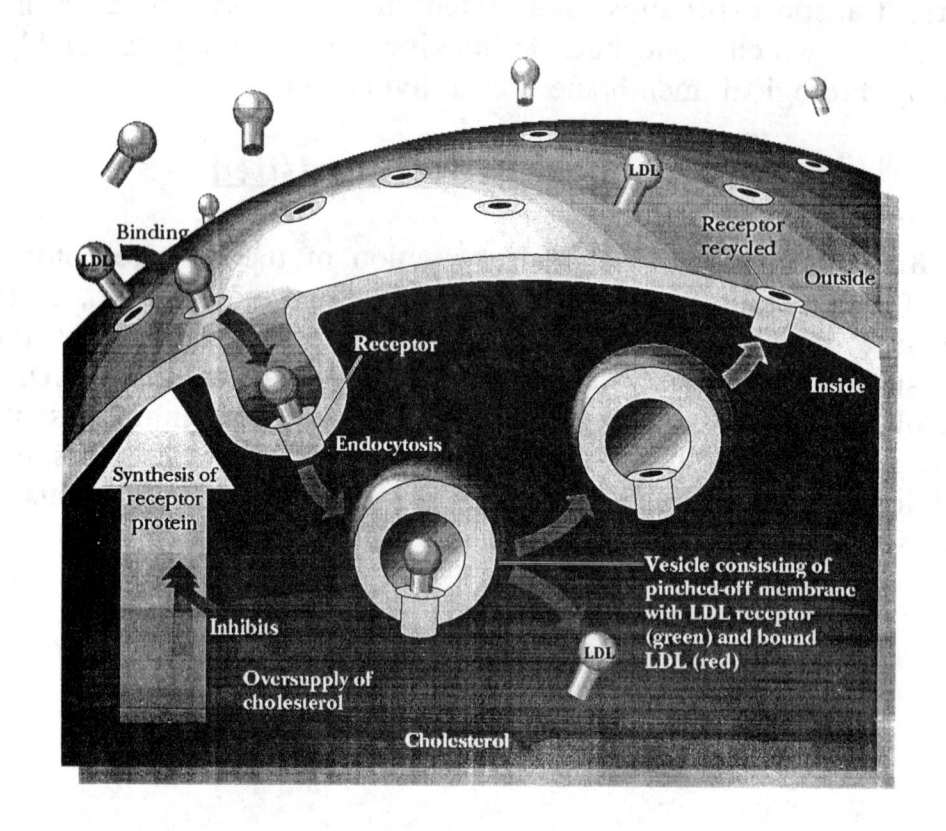

Chapter 8 Lipids and Membranes

CHAPTER OBJECTIVES

1. State the criterion for a molecule to be classified as a lipid.
2. Write the general equation for the reaction of:
   a. a carboxylic acid and glycerol to form an ester.
   b. a sphingosine and phosphoric acid.
   c. a phosphoric acid and an alcohol.
3. Distinguish between a triglyceride, phosphatidic acid, a phosphoacylglyceride and a glycolipid.
4. Point out the common linkages found in molecules such as phosphatidylethanolamine, phosphatidylcholine and phosphatidylinositol.
5. Draw the structure for the general steroid skeleton and name three important molecules derived from this basic structure.
6. Draw the structures for a glycolipid and a phosphoglyceride and indicate why they are considered amphiphilic.
7. Describe the lipid bilayer model for a membrane.
8. Name and draw the molecules that make up the membranes of animal cells and indicate their relative orientation within the bilayer.
9. Indicate the factors contributing to the fluidity of a lipid bilayer or a biological membrane.
10. Describe how a cooperative structural transition (ordered to disordered) occurs when thermal energy is added to a membrane.
11. Explain the difference between lateral and "flip-flop" or transverse diffusion of lipids in a membrane and indicate the relative probability of each.
12. Outline the characteristics of peripheral and integral proteins and indicate some important functions they perform in the membrane.
13. Describe the fluid mosaic model for a biological membrane.
14. Distinguish between passive, facilitated and active transport mechanisms for the movement of molecules and ions from one side of the membrane to the other.
15. Generally characterize a membrane receptor and give an example of how a membrane receptor relates to a specific medical problem.
16. Describe the difference between a ligand-gated and a voltage-gated channel and explain how they may be involved in a neuromuscular junction.
17. Examine the structures of the fat soluble vitamins, point out any functional groups and give two functions for these essential molecules.

Chapter 8 Lipids and Membranes

18. Compare the structure of a prostaglandin and a leukotriene molecule with their precursor molecule, arachidonic acid.

SPOTLIGHT

Controversial Drugs: Synthetic Steroids

It is now common knowledge that many professional, and even amateur athletes, take androgenic-anabolic steroids to increase muscle mass and strength. These steroids are synthetic male hormones which were originally synthesized to promote tissue-building (anabolic) effects and to eliminate or reduce their androgenic (masculizing) effects.

In the 1988 Summer Olympic games that were held in Seoul, Korea, a Canadian sprinter, Ben Johnson, ran the 100-meter race in 9.79 seconds to win the gold medal. Drug tests revealed that Johnson had significant levels of the anabolic steroid, stanozolol, in his urine. Since the use of anabolic steroids is illegal by Olympic rules, Johnson was stripped of his victory and the Olympic gold medal.

The structures for the female and male sex hormones, estradiol and testosterone, respectively, are shown below and they are compared to the synthetic anabolic steroid, stanozolol. Common features of male hormones include a methyl group at the carbon atom, which serves as the "top" junction for the two six-membered rings in the left of the drawing, and the absence of an aromatic ring in the steroid structure. A five-membered heterocyclic ring has also been added on to make the stanozolol molecule. Stanozolol is but one of the many synthetic anabolic steroids used by athletes today, despite the many detrimental side-effects which have been reported.

Another controversial synthetic steroid, referred to as RU-486, is a drug developed in France in 1980 to terminate pregnancy. The structure of RU-486 (shown below) is similar to progesterone and it binds to progesterone receptors on the surface of the cell. To the surprise of the biomedical researchers, the binding of RU-486 to the receptors does not

Chapter 8 Lipids and Membranes

produce the same hormonal response within the cell as that produced by progesterone (molecules that exhibit the same type of behavior with receptors are generally called <u>agonists</u>). Instead, the binding to the progesterone receptor blocked any response (molecules of this type are referred to as <u>antagonists</u>). RU-486 is a progesterone antagonist and not only blocks progesterone binding, but also blocks its physiological activity. This was a key finding, especially in light of well established results which indicated that progesterone binding to its receptor was essential for the progression of a successful pregnancy. RU-486 is approved for use in abortions in France. Although many successes have been reported, the details of the mechanism of action of this drug and its possible short-term and long-term side effects remain to be elucidated.

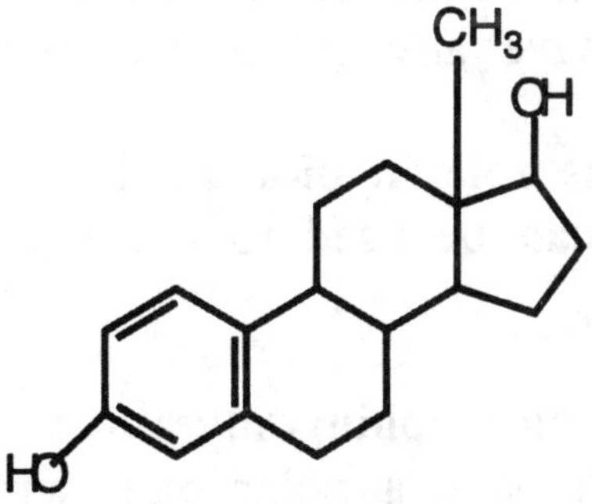

Estradiol

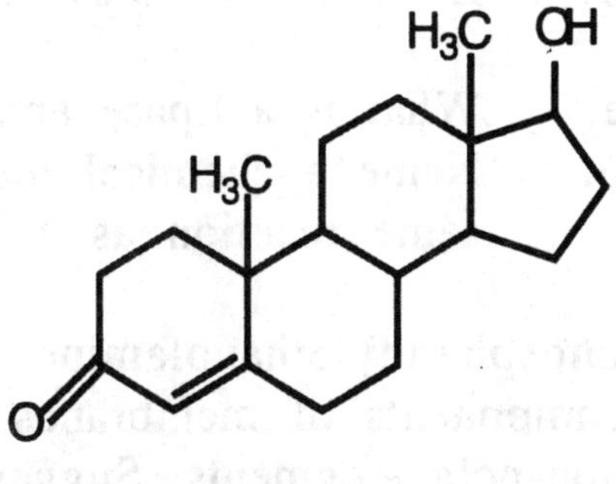

Testosterone

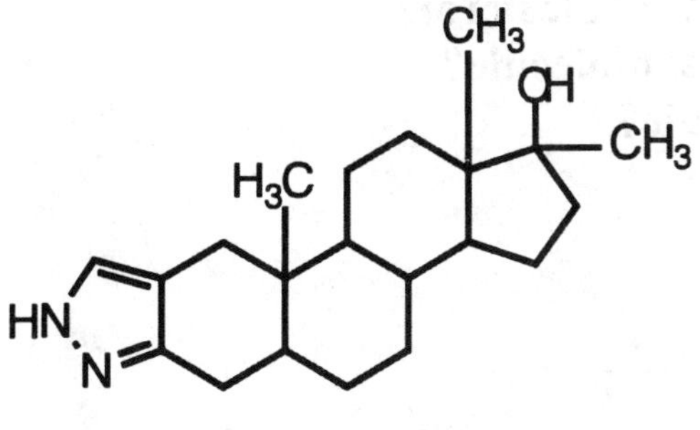

Stanozolol

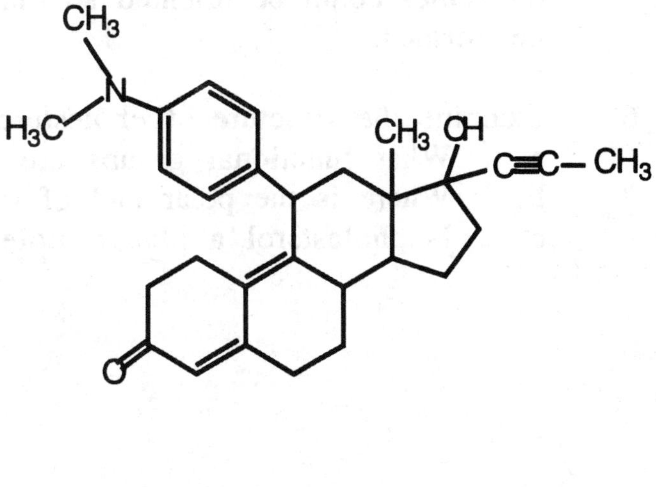

RU-486

Chapter 8 Lipids and Membranes

EXERCISES

1. a. What is the definition of a lipid?
   b. How does the manner in which one defines a lipid differ from that for carbohydrates, proteins and the other important biomolecules?

2. a. List a number of outstanding general characteristics of fatty acids found in organisms.
   b. Are the double bonds in the unsaturated fatty acids conjugated or isolated?
   c. At which end of the fatty acid does the numbering of the carbons begin?
   d. In four of the five examples of unsaturated fatty acids in Table 8.2, the double bond nearest the acid group is formed between which two carbon atoms? What are the names of t these acids? Which acid in the Table is an exception to this observation?

3. Two molecules serve as the backbone for many lipids, including phosphoglycerides and ceramides.
   a. Name the two molecules that serve as the backbone.
   b. Do either of these molecules have any chiral centers?

4. a. What is a lipase and what is its role in biological systems?
   b. Name a chemical method that can be used to accomplish the same reaction as a lipase?

5. Phosphatidylethanolamine (cephalin) and sphingomyelin are both components of membranes. They both have a polar end, with long non-polar segments. Suggest a way in which both of these molecules could be oriented so that a micelle or a lipid bilayer could be formed.

6. Examine the structure of cholesterol in Figure 8.7 in the text.
   a. What functional groups are in cholesterol?
   b. Where is the polar end of this molecule?
   c. Is cholesterol a planar molecule?

Chapter 8 Lipids and Membranes

7. The melting points for two 16 carbon fatty acids is given below. Palmitic acid is completely saturated, while there is one double bond in palmitoleic acid. Explain the molecular basis for the trend in melting points.

Fatty Acid		M.P.(°C)
Palmitic acid	$CH_3(CH_2)_{14}COOH$	63.1
Palmitoleic acid	$CH_3(CH_2)_5CH=CH(CH_2)_7COOH$	-0.5

8. A five carbon unit called isoprene is found as a reoccurring or often repeating unit in a number of lipids, including some vitamins and/or their precursors. The structure of the unit is:

   $$CH_2=C-C=CH_2$$
   with H on the second carbon and CH_3 on the third carbon.

   Examine the structures of β–carotene, vitamin A, D and K, arachidonic acid, prostaglandins and leukotrienes found in the text. Does the isoprene unit occur in these molecules?

9. Match the fat soluble vitamin in the left-hand column with its characteristic in the right column.

   Vitamin D   a. _____ retinol; involved in vision
   Vitamin E   b. _____ regulation of calcium and phosphorus metabolism
   Vitamin K   c. _____ contains a bicyclic ring with two carbonyls (a quinone); involved in blood coagulation
   Vitamin A   d. _____ antioxidant; reacts with free radicals; active form is α-tocopherol

10. Acquired immune deficiency syndrome (AIDS) is a disease which is caused by the viral infection of cells associated with the development of an immune response. The so-called helper T cells, involved in cellular immunity, are the ones that are specifically

Chapter 8 Lipids and Membranes

infected. Suggest a reason for the apparent specificity of the HIV (Human Immunodeficiency Virus) for these T cells.

11. Consider the accompanying composite drawing of the cell, which shows molecules or ions that must be transported into or out of the cell or one of the organelles. For each small molecule, indicate whether the transport process involves:
   a. passive transport
   b. facilitated transport
   c. active transport

   In each case, indicate whether the process is a spontaneous one (i.e., a decrease in the free energy, $-\Delta G$, for the process).

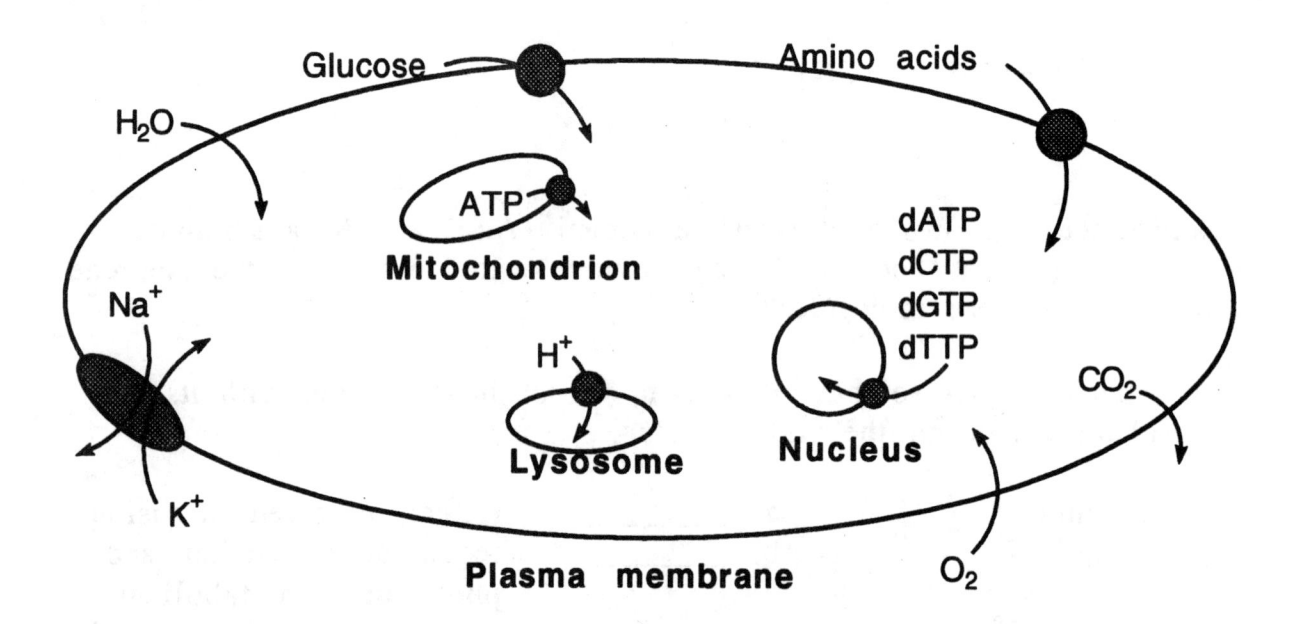

A Collection of a Few Transport Systems in Most Cells

12. The free energy change that occurs when an uncharged species at concentration, c_1, is transported from side 1 to side 2, where it is at concentration, c_2, is given by the equation:

   $$\Delta G = 2.303\, RT \log [c_2]/[c_1]$$

   a. If the concentration of $c_1 > c_2$, is the process spontaneous?
   b. If the concentration of $c_1 < c_2$, is the process spontaneous?

Chapter 8 Lipids and Membranes

c. Consider the situation in which the glucose concentration outside the cell is 20 mM and that inside the cell is 20 uM (micromolar). What is the free energy change for the transport of glucose into the cell at 37°C?

13. The kinetic profile for the transport of ethanol (a) and glucose (b) across a cell membrane is shown in the following Figure. Explain these profiles in terms of the mechanisms by which these two molecules are transported across a membrane. The concentrations of the molecules outside and inside the cell are c_1 and c_2, respectively.

a.

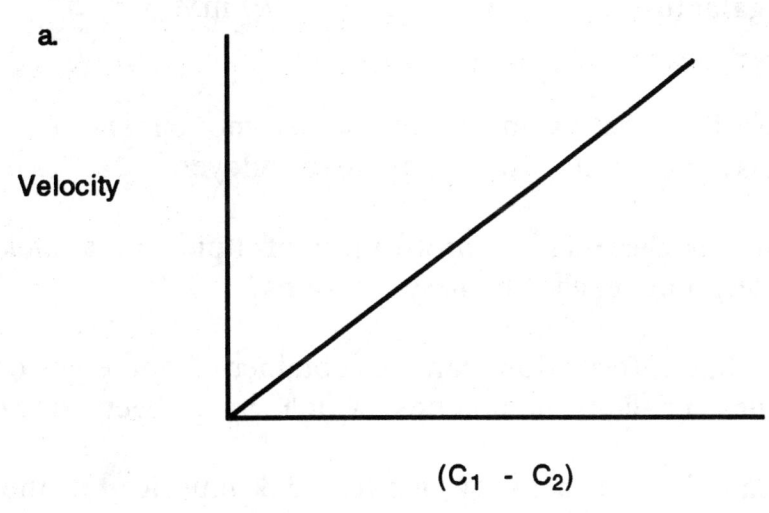

Velocity vs $(C_1 - C_2)$

b.

Velocity vs $(C_1 - C_2)$

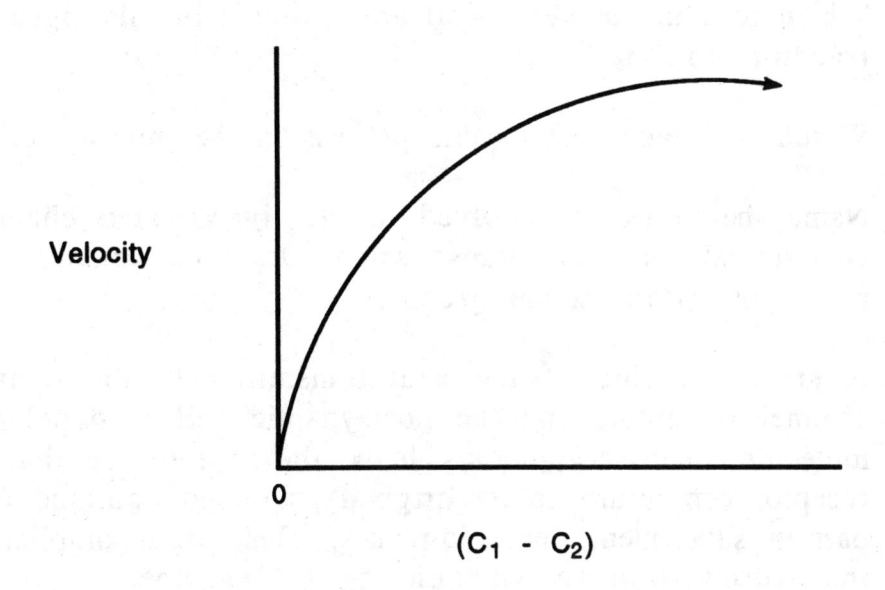

Chapter 8 Lipids and Membranes

14. The data in the Table below show the K_M values for the transport of a number of sugars by the erythrocyte glucose transport protein across a red blood cell membrane.

    a. What is the definition of the K_M value?

    b. Interpret these findings.

Sugar	K_M Value
D-glucose	1.5 mM
L-glucose	>3000 mM
D-mannose	20 mM
D-galactose	30 mM

15. a. Describe what is meant by lateral motion and the "flip-flop" or transverse motion of lipids in a bilayer.

    b. What is the relative probability of lipids in a membrane undergoing each of these motions?

16. Indicate what information can be obtained from electron micrographs of cell membranes which have been freeze-fractured.

17. The communication between nerves and muscles at the neuromuscular junction involves both voltage- and ligand-gated channels.

    a. Which ions in the nerve cell are involved in this signal transmission?

    b. Which ions take part in this process in the muscle cell?

    c. Name the molecule involved in the ligand-gated channel which is essential for nerve transmission. Draw its structure and point out its functional groups.

    d. After the binding of the neurotransmitter to the ligand-gated channel (receptor) and the postsynaptic cell is depolarized, the molecule must very quickly leave the receptor so that the receptor can return to its original state and continue taking part in subsequent nerve impulses. This is accomplished by the hydrolysis of acetylcholine by the enzyme, acetylcholinesterase. At which bond does this enzyme act and what are the resulting products?

Chapter 8 Lipids and Membranes

e. The molecule, succinylcholine, shown below, is used as a muscle relaxant. Suggest a reason for its effectiveness.

$$(CH_3)_3\overset{+}{N}-CH_2-CH_2-O-\underset{\underset{O}{\|}}{C}-CH_2-CH_2-\underset{\underset{O}{\|}}{C}-O-CH_2-CH_2-\overset{+}{N}(CH_3)_3$$

ANSWERS TO EXERCISES

1. a. Lipids are a family of biomolecules which are soluble in non-polar organic solvents, such as ether or chloroform.

   b. Lipids are defined operationally and not by their composition or their function; therefore, lipid molecules exhibit a wide variety of functions. The lipid family includes the phospholipids and glycolipids, which are major components of the lipid bilayer, of membranes, of the steroid hormones, of energy stores (e.g., fats), and of fat soluble vitamins.

2. a. Fatty acids presented in the chapter have the following characteristics:

   They can be saturated or unsaturated.
   They contain an even number of carbon atoms.
   There are no branched fatty acids.*
   They are amphiphilic in character.

   For unsaturated fatty acids:
   The stereochemistry is *cis* in virtually all cases.
   With few exceptions, the double bonds are isolated and not in a conjugated system.

   b. For double bonds to be conjugated, the double bonds must be on adjacent carbons. If the double bonds are on non-adjacent carbons, as in the fatty acids, they are non-conjugated or isolated double bonds.

Chapter 8 Lipids and Membranes

Double Bond Systems

conjugated
(double bonds on adjacent carbons)

$$\overset{*}{-CH_2} = CH - \overset{*}{CH} = CH_2 -$$

non-conjugated
(double bonds on non-adjacent carbons)

$$\overset{*}{-CH} = CH - CH_2 - \overset{*}{CH} = CH-$$

    c.   The numbering of the carbon atoms starts at the carboxylic acid carbon.

    d.   The double bond closest to the acid functional group is between C_9 and C_{10}.

This occurs commonly and is found in palmitoleic (C_{16}), oleic (C_{18}), linoleic (C_{18}) and linolenic (C_{18}) acids. The latter two fatty acids are polyunsaturated. Arachidonic acid is an exception to this general observation.

3.   The two molecules which serve as the backbone in these lipids are glycerol and sphingosine. Although glycerol has no chiral centers, sphingosine has two chiral carbons. Examine Figure 8.2 and 8.6 (in the text) to locate the chiral carbons.

4.   a.   A lipase is an enzyme which catalyzes the cleavage of ester linkages, such as in triglycerides, to form glycerol and fatty acids.

    b.   Acidic or basic (called saponification) hydrolysis is a chemical route to accomplish ester hydrolysis. Saponification is the procedure used to produce soaps (Figure 8.3 in the text).

5.   next page

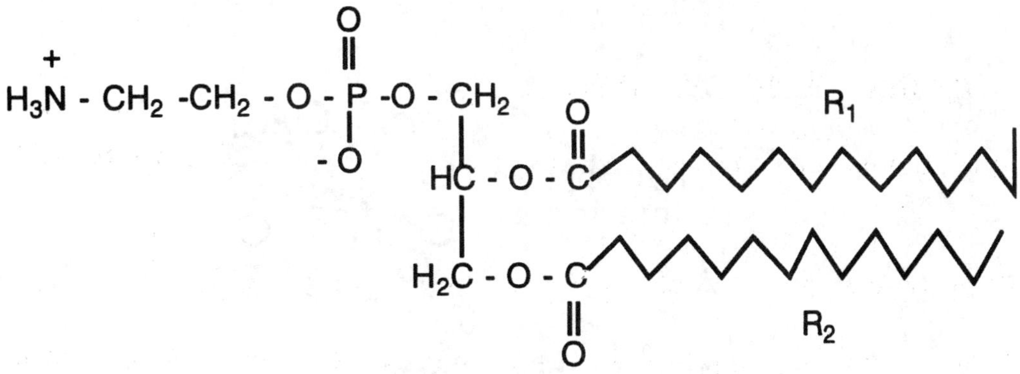

Phosphatidylethanolamine

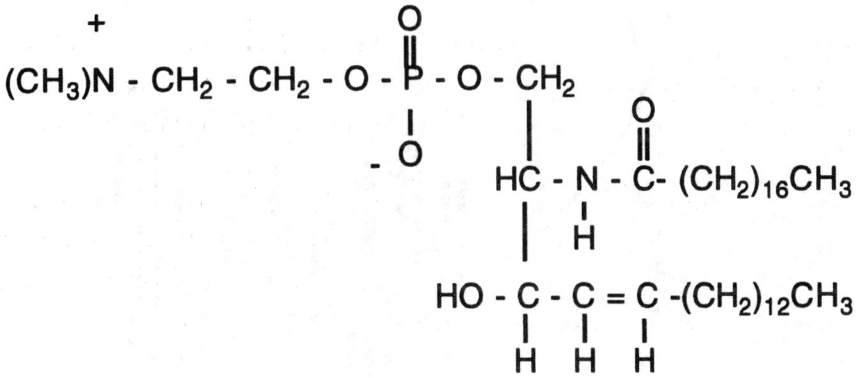

Sphingomyelin

In both cases, the amphiphilic structures can be generally drawn as shown below.

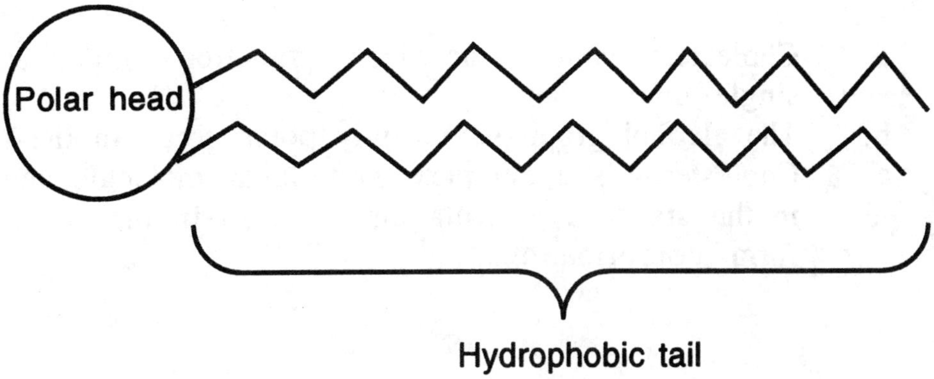
Polar head
Hydrophobic tail

Chapter 8 Lipids and Membranes

Hydrophobic Interior

Hydrophilic, Charged Exterior

A Micelle

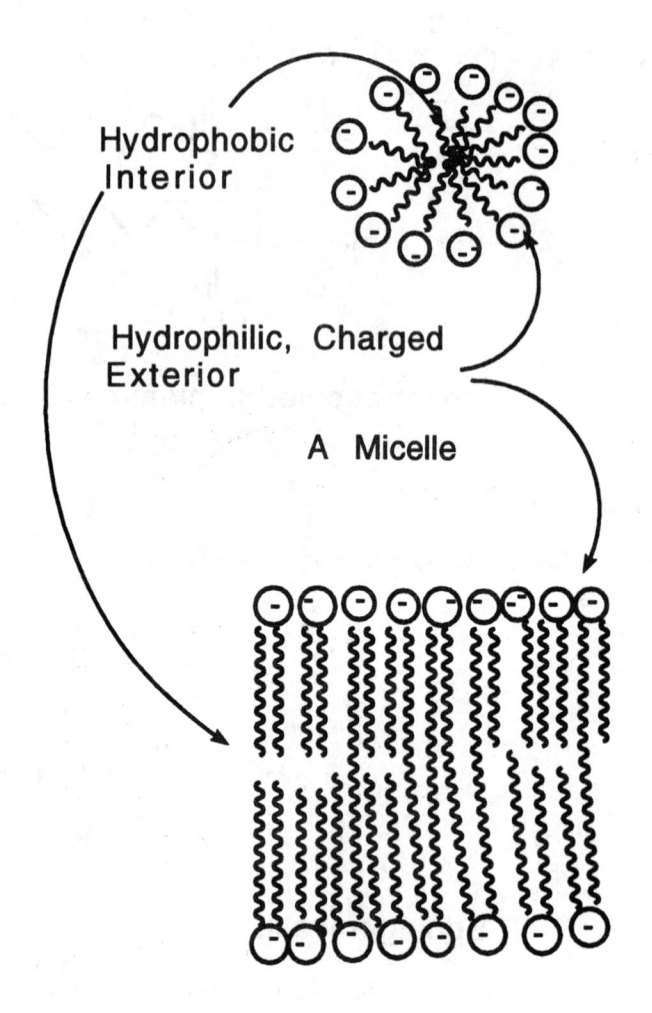

Lipid bilayer

Both the micelle and the lipid bilayer structures can form spontaneously in aqueous solution. The hydrophilic polar heads interact favorably with the aqueous solution, while the hydrophobic tails associate favorably with each other and are not exposed to the aqueous solution.

6.   a.   Cholesterol contains an alcohol functional group and has a single double bond.
     b.   The alcohol group is the only polar group in the molecule.
     c.   Cholesterol is a complex, non-planar molecule shown below in the standard presentation and also in the "boat and chair" form conformations.

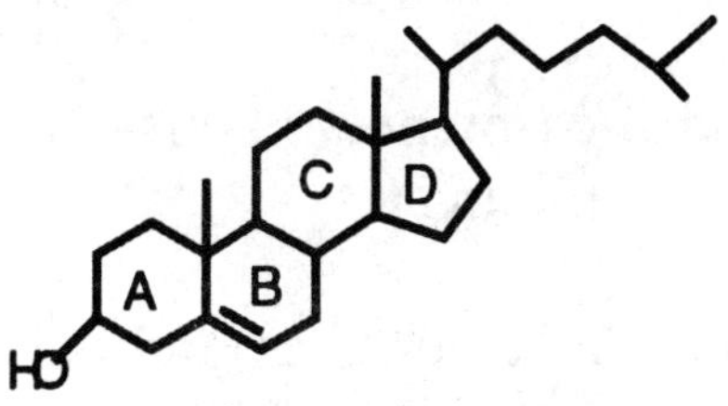

Chapter 8 Lipids and Membranes

Cholesterol

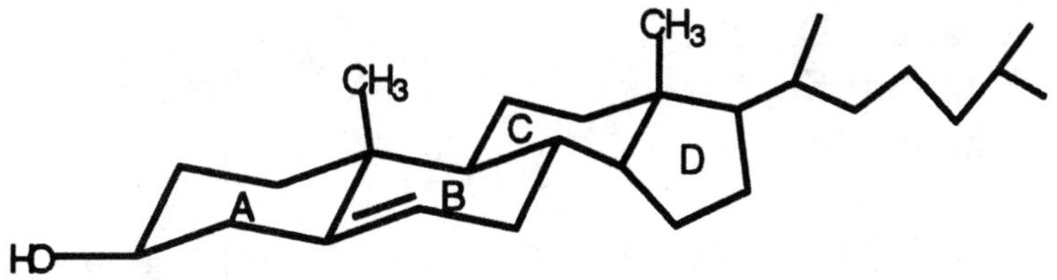

7. The melting point is a function of the **inter**molecular forces between the molecules. In the case of the fatty acids, this involves hydrophobic interactions between the long organic groups. For palmitic acid, the degree of alignment of the saturated chains in the fatty acid is very high, resulting in a high degree of interaction between the molecules. This results in a relatively high melting point.

On the other hand, with palmitoleic acid, there are a larger number of possible (unfavorable) orientations that an unsaturated organic chain can take. This results in an overall lower degree of alignment and therefore the number of similar interactions between the molecules is decreased. The following figure generally illustrates this.

High degree of alignment in saturated palmitic acid

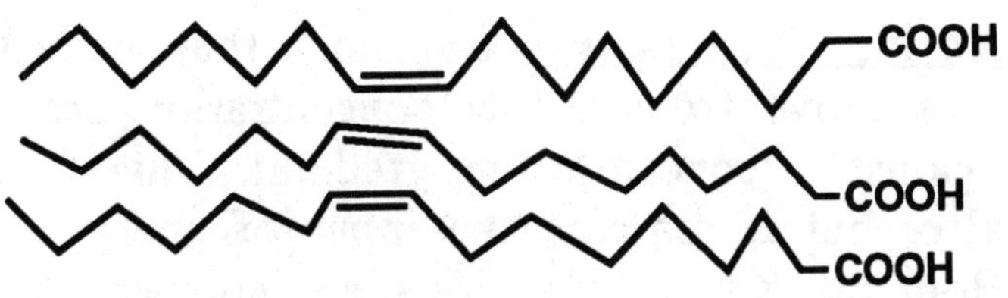

Degree of alignment is decreased in unsaturated palmitoleic acid

Chapter 8 Lipids and Membranes

8. The isoprene unit is found in β-carotene, 11-cis-retinal, its derivatives (Figure 8.25) and vitamin K.

   Vitamin D and E, arachidonic acid, prostaglandins and leukotrienes (Figures 8.27 and 8.30 and the unlabeled figure after 8.27) do not contain the isoprene units.

9. a. Vitamin A
   b. Vitamin D
   c. Vitamin K
   d. Vitamin E

10. The specificity of the HIV virus for T cells is due to the presence of receptor proteins, called CD 4 receptors, on the surface of the T cells. The first step in the infection is the binding of the virus to the cell surface. This binding occurs through the specific interaction of the HIV virus with the CD 4 receptor, a large transmembrane protein.

11. The Figure shows a variety of molecules which are transported into or out of the cell or an organelle. The transport of the ions and molecules can be cataloged accordingly;

Passive Transport - random and simple diffusion through a lipid bilayer membrane, primarily involving H_2O and small non-polar molecules. This includes the molecules, CO_2 and O_2. This transport is a spontaneous process involving no input of energy.

Facilitated Transport - transport that occurs only with the assistance of a specific membrane carrier protein, often called a permease or translocator protein. The molecules associated with this type of transport are glucose and amino acids. In these cases, the molecules diffuse from a compartment of high concentration to one of a lower concentration. Therefore, there is no expended energy and the transport is spontaneous.

Active Transport - transport that occurs when specific molecules or ions move from a low concentration area to high concentration area, against a concentration gradient. This is a facilitated transport process also, but it requires the input of energy. This is not a spontaneous process. Both the K^+ and Na^+ ions are pumped into and out of the cell, respectively, against a concentration gradient by the Na^+-K^+ pump. This multisubunit transmembrane protein, referred to as the Na^+-K^+ ATPase, links the transport of Na^+ and K^+ ions, in that it pumps three Na^+ ions out, for every

Chapter 8 Lipids and Membranes

two K^+ ions pumped into the cell. This occurs with the concomitant hydrolysis of ATP to provide the energy to make this an exergonic process.

The H^+ also undergoes active transport into the lysosomes since its concentration is much higher in the lysosomes than in the cytoplasm.

In the case of glucose transport, in which there is transport of only a single molecule across a membrane, the mediated transport process is called **UNIPORT**. In the case of the Na^+-K^+ ATPase, the transport is referred to as an **ANTIPORT**, because two different molecules or ions are simultaneously transported in opposite directions.

For the ATP transport (an anionic species) out of the mitochondrion, a transport protein has been isolated, but the process is not well characterized at this time. Whether it involves a facilitated transport requiring energy is yet to be resolved.

The mechanism of transport of the deoxyribonucleotides (dATP, etc) into the nucleus is likewise not firmly understood at this time. An interesting feature of the nuclear membrane is that it has relatively huge "pores" which permit the transport of large molecules into and out of the nucleus.

Chapter 8 Lipids and Membranes

Because of the unusual size of these pores, it is quite possible, but not certain, that passive transport may be involved in this process.

These last two cases emphasize that biochemistry is an ongoing venture in which many important questions are still being addressed and answers are being earnestly sought. Our efforts are only beginning to give us a more comprehensive understanding of the functioning of the cell.

12. a. Since this process involves the transport of a species from a side at higher concentration to one at a lower concentration, the process is spontaneous and the ΔG will be negative.
    b. The situation is just the opposite when $c_2 > c_1$. In this case, the process would involve the transport of a species from a side with a lower concentration to the side with a higher concentration. This is not a spontaneous process and ΔG will be positive. This process will only occur if there is an input of useful energy from another source.
    c.
    $\Delta G = 2.303\ RT \log [C_{in}]/[C_{out}]$
    $\Delta G = 2.303\ (8.31\ J/mol\text{-}K)\ (310K) \log [20\ uM/\ 20,000\ uM]$

    $\underline{\Delta G\ =\ -\ 17.8\ kJ/mol}$ This is a spontaneous process.

13. The profiles indicate that the mechanism of transport is different for the two molecules. The kinetic characteristics for the transport of ethanol is a classic profile for **simple diffusion**. The greater the concentration gradient across the membrane, the greater will be the diffusion rate.

    On the other hand, the transport of glucose exhibits quite different characteristics. This profile is analogous to that of Michaelis-Menten kinetics for reactions catalyzed by simple (non-allosteric) enzymes. Recall that the Michaelis-Menten kinetic model assumes the interaction of a substrate with the enzyme to produce the enzyme-substrate complex, ES, which then dissociates to product and enzyme.

    $$S + E \rightleftharpoons ES \longrightarrow product + E$$

This profile then suggests an analogous process for glucose transport in which glucose must interact with a transport protein to facilitate its transport across the membrane. At low concentrations of glucose, the

Chapter 8 Lipids and Membranes

kinetics are linear (i. e., first order kinetics) since there is an abundant supply of transport protein to readily facilitate transport of all the glucose. However, as the glucose concentration increases, the amount of transport protein becomes comparable to that of glucose and the kinetics are no longer strictly linear. At very high concentrations, there is no change in the transport velocity as the glucose concentration increases (i.e., zero order kinetics). This indicates that the transport protein is limiting and the protein that is there, is completely saturated. Note that a value for V_{max} and K_M can be obtained for the transport process by an analysis which utilizes a Lineweaver-Burk plot.

This is the characteristic profile found for cases in which the transport is by **facilitated diffusion.** One final point; although glucose can be transported through a membrane by simple diffusion also, the mechanism of facilitated diffusion predominates because it is so much more efficient. For glucose, facilitated diffusion provides a transport mechanism which is about a factor of 500 greater than that for simple diffusion.

Chapter 8 Lipids and Membranes

a.

[Graph: Velocity vs (C_1-C_2), linear increasing line]

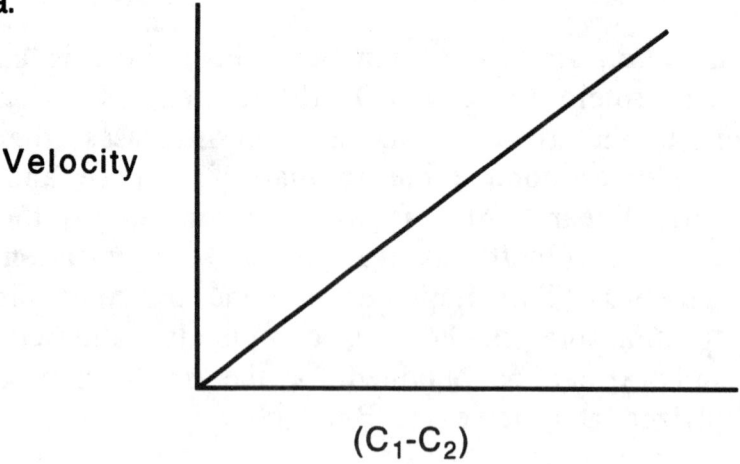

b.

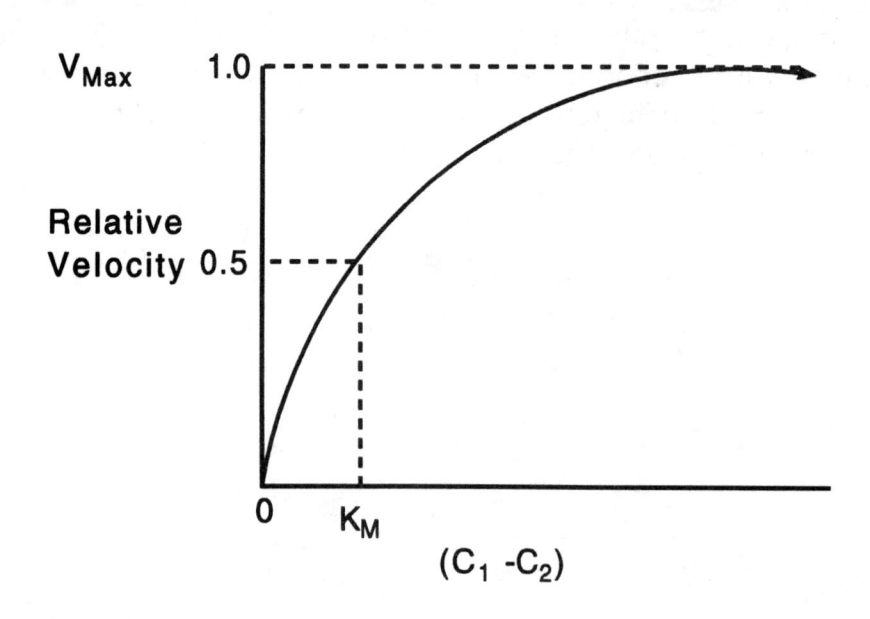

[Graph: Relative Velocity vs (C_1-C_2), saturation curve approaching $V_{Max} = 1.0$, with K_M marked at relative velocity 0.5]

Plots of initial rates of diffusion versus concentration difference

14. a. The K_M value denotes the concentration at which the transport velocity occurs at one-half the maximum rate. It is an expression of the "goodness-of-fit" of the substrate for the permease or transport protein, analogous to the K_M value in enzyme kinetics. The LOWER the value for K_M, the better the fit and the more proficient the transport process.

  b. The data indicate that D-glucose has the lowest K_M value of the four sugars studied. The value is many orders of magnitude lower than that for the L form of glucose (an enantiomer) and

Chapter 8 Lipids and Membranes

more than one order of magnitude over that for either mannose or galactose. These latter two monosaccharides are epimers of glucose (they differ in the configuration at only one asymmetric center). These data therefore emphasize the important point that the transport protein for glucose is highly specific for the molecule it transports. This is similar to the characteristics of enzymes, in which the enzymes are highly specific for the substrates upon which they act. This finding of specificity for a molecule can be regarded as a general characteristic of other transport proteins.

15. a. The lipid bilayer is often referred to as being made up of two "leaflets" of phospholipid molecules.

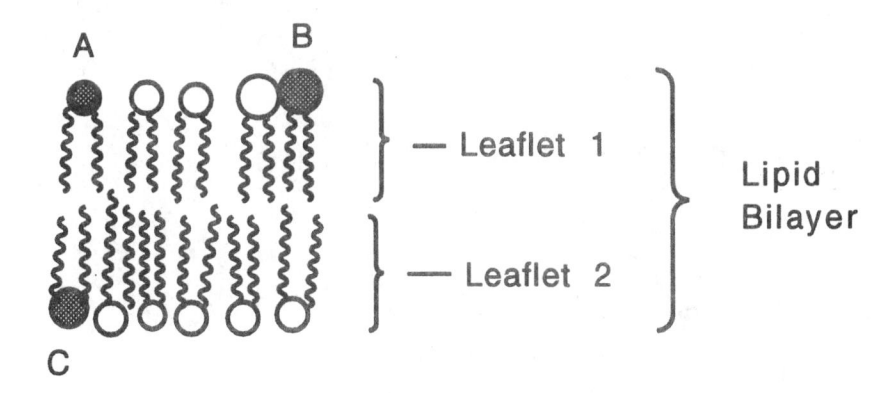

    Lateral motion is the movement of a phospholipid from one region of a leaflet to another region in the **same leaflet** (the movement of the molecule in position A to position B). On the other hand, traverse or "flip-flop" motion involves the **transfer of a phospholipid from one leaflet to the other leaflet** (the movement of the molecule at position A to position C).

    b. The probability of these two different motions is very different because of the energetics involved in each process. Lateral movement is energetically favorable and therefore very rapid (occurs continuously). It involves the movement of the molecules in their same orientation, keeping the polar head groups on the outside of the bilayer, interacting favorably with the aqueous environment, while the non-polar tails remain in the interior, interacting with other non-polar tails.

Chapter 8 Lipids and Membranes

Transverse or "flip-flop" diffusion of a phospholipid molecule requires the movement of the polar head group through the non-polar regions. This is energetically unfavorable and therefore a relatively rare event.

16. Freeze-fracture of a cell membrane opens up and separates the membrane such that the inner side of the leaflets (the non-polar tails) can be examined. A granular appearance indicates that there are integral proteins in the membrane.

17. a. The ions in the nerve cells are Ca^{+2}.
    b. The ions in the muscle cells are Na^+ and Ca^{+2}.
    c. Acetylcholine, which contains both an ester and an (quaternary) amine functionality, is shown below.

$$CH_3-C(=O)-O-CH_2-CH_2-N^+(CH_3)_3$$

d.

Ester Linkage

$$CH_3-C(=O)-O-CH_2CH_2-N^+(CH_3)_3 \xrightarrow{AC} CH_3-C(=O)-OH + HO-CH_2CH_2-N^+(CH_3)_3$$

Acetic acid   Choline

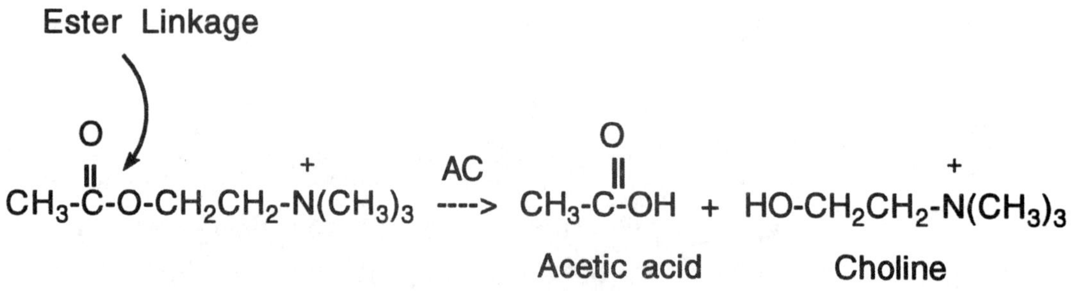

AC = acetylcholinesterase

e. Succinylcholine is a structural analog of acetylcholine, which is a competitive inhibitor of acetylcholinesterase. It competes for the active site, and although it is hydrolyzed by acetylcholinesterase, this occurs at a much slower rate than acetylcholine. This results in a longer depolarization and muscle contraction is inhibited.

9

The Importance of Energy Changes and Electron Transfer in Metabolism

Every reaction has an intrinsic driving force, which determines the extent to which it will proceed to completion. As a reaction proceeds and ultimately attains equilibrium, the **entropy** (S) of the universe (system plus the surroundings) will be a maximum, while the **free energy** (G) for the molecules (the system) reaches a minimum. In this respect, $\Delta S_{univ} > 0$ and $\Delta G_{sys} < 0$ are fundamental indicators of the spontaneity of any reaction. The **standard free energy**, ΔG^o, is the free energy difference between the products and the reactants under standard conditions (i.e, 1 M for all solutes and 1 atm for gases at 25°C). The equilibrium constant, K_{eq}, and the ΔG^o are really two different means to express this intrinsic driving force. Since most important biochemical reactions are carried out near neutral pH, a standard free energy, $\Delta G^{o'}$, and the equilibrium constant, K'_{eq}, are defined similarly, except that the standard state concentration for hydrogen ions [H^+] is 1×10^{-7} M. The free

Chapter 9 The Importance of Energy Changes and Electron Transfer in Metabolism

energy change of a reaction is also related to the **enthalpy change**, ΔH, for the reacting system and the entropy change of the universe. It is essential to realize that thermodynamic quantities such as G, H and S are **state functions**. The important property of any of these functions is that the difference in two states [e. g., $\Delta S = (S_{products} - S_{reactants})$] depends only on the initial and final states and not on the route by which the reaction
proceeds.

Metabolism can generally be considered as the breakdown of foods that are ingested to provide energy and the fundamental building blocks for the (bio)synthesis of biomolecules required to sustain life. Although many types of reactions are involved, **catabolism** is an oxidative process which liberates energy, while **anabolism** is a reductive process which requires the input of energy. Coenzymes, such as NADH, NADPH, $FADH_2$ and their oxidized forms, serve as electron donors and acceptors. Together with molecules such as ATP, which contain "high energy" bonds, these small molecules act as essential partners in many of the enzymatic reactions. In many metabolic pathways, endergonic reactions are coupled to exergonic reactions, with the latter providing the thermodynamic driving force for a thermodynamically favored, coupled reaction. The essential role of ATP and molecules taking part in electron transfer will become vividly clear as the processes of glycolysis, the citric acid cycle and oxidative phosphorylation are explored in subsequent chapters.

Some Key Figures Revisited

Figure 9.7 A comparison of catabolism and anabolism.

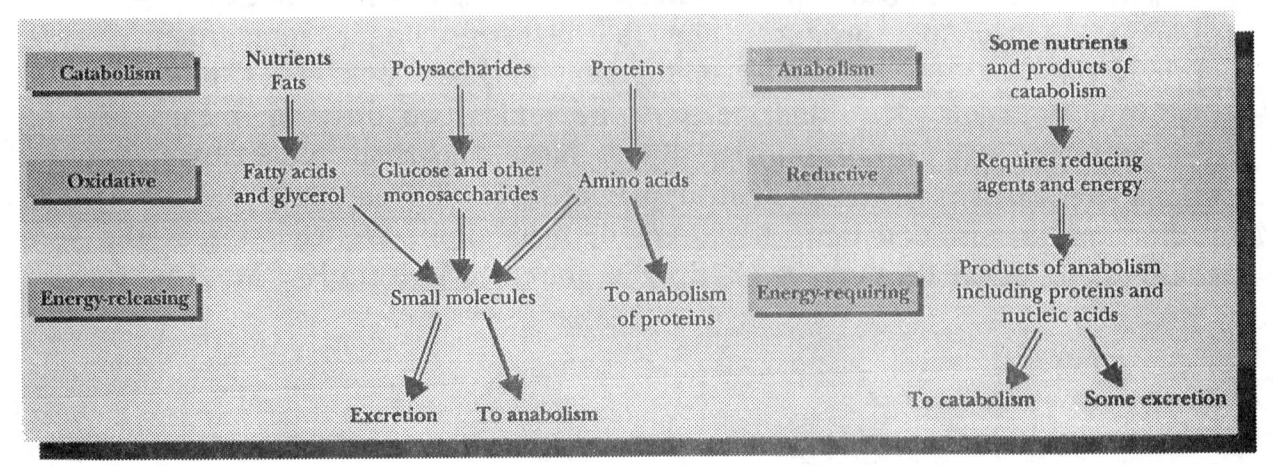

Chapter 9 The Importance of Energy Changes and Electron Transfer in Metabolism

Figure 9.13 The role of electron transfer and ATP production in metabolism. NAD$^+$, FAD, and ATP are constantly recycled.

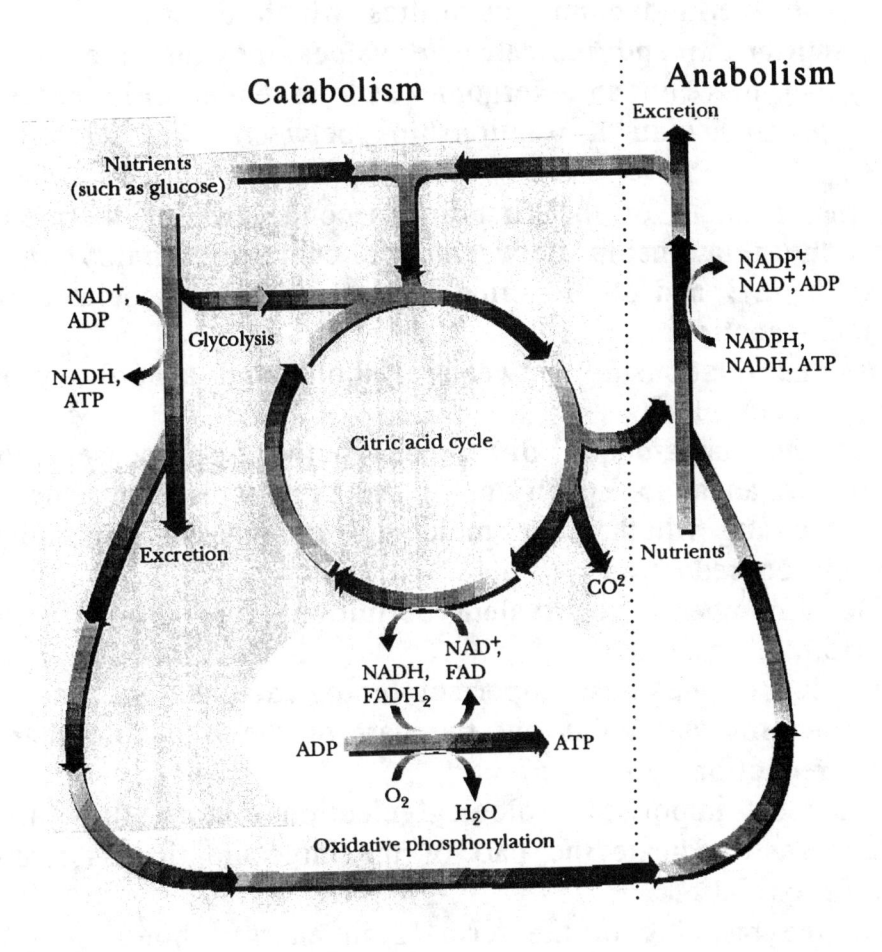

Chapter 9 The Importance of Energy Changes and Electron Transfer in Metabolism

CHAPTER OBJECTIVES

1. State the thermodynamic criteria for the spontaneity of a reaction.
2. Define the thermodynamic quantities which determine if a reaction is at equilibrium and indicate the values they assume.
3. Distinguish between an exergonic and an endergonic process.
4. Write the mathematical relationships between $\Delta G'$, $\Delta G^{o'}$ and K'_{eq}.
5. State the meaning of the first and second laws of thermodynamics.
6. Define the relationship between free energy, enthalpy and entropy and define ΔH and ΔS in terms that are readily associated with a chemical reaction.
7. Discuss the relationship between catabolic and anabolic processes in cellular metabolism.
8. Indicate the fundamental differences between processes which are catabolic or anabolic, specifying if the process is exergonic or endergonic and whether the molecules of interest are undergoing oxidation or reduction.
9. Provide a number of equivalent definitions for both oxidation and reduction.
10. Make a list of the most important biological electron acceptors. Draw their structures and point out the part of the molecule that is altered by the reduction.
11. List the most important biological electron donors; draw their structures and indicate the part of the molecule that is altered as a result of oxidation.
12. Explain the meaning of the term "high energy" bond.
13. Draw the structure for ATP. Point out the acid anhydride and the ester linkages and indicate the high energy bonds.
14. Indicate the number of electrons and protons that NADH, NADPH and $FADH_2$ transfer when they are oxidized.
15. Discuss how the coupling of an endergonic reaction to an exergonic reaction can produce an overall reaction which is thermodynamically favored.
16. Draw the structure for coenzyme A (CoA). Point out the recognizable components in its chemical structure and indicate the role of CoA in metabolism.
17. List the three important processes in the metabolism of glucose and indicate whether they are anaerobic or aerobic.

Chapter 9 The Importance of Energy Changes and Electron Transfer in Metabolism

EXERCISES

1. Indicate which of the following statements are true or false. If a statement is false, correct it.
   a. The first law of thermodynamics states that energy can neither be created nor destroyed.
   b. The second law of thermodynamics states that the entropy of the universe is constantly increasing.
   c. The free energy change for a reaction at equilibrium is zero.
   d. The standard free energy change for a reaction at equilibrium is zero.
   e. The enthalpy change of a chemical reaction is derived primarily from the energies of bonds that are broken and formed in the reaction.
   f. The burning of a piece of paper is an example of a reaction that is spontaneous with increasing entropy.
   g. A reaction which is spontaneous under one particular condition will be spontaneous under all other conditions.
   h. A calorie is greater than a joule of energy.
   i. Thermodynamics indicates whether or not a reaction will proceed spontaneously.
   j. Thermodynamics can be used to predict how fast a reaction takes place.
   k. The mixing of oil and water produces an increase in entropy.
   l. Hydrophobic interactions are spontaneous because of favorable values for both the enthalpy and entropy change.
   m. Endergonic reactions have a positive free energy change and are not spontaneous.
   n. Endergonic reactions can occur if free energy is supplied from another source, such as the coupling with another reaction.

2. The K_{eq} for the hydrolysis of adenosine monophosphate (AMP) to adenosine (A) and phosphate (P_i) at 25°C is $3.0 \times 10^{+2}$.
   a. What are the values of $\Delta G'$ and $\Delta G^{o'}$ for this reaction at equilibrium?
   b. Compare the $\Delta G^{o'}$ for this reaction with that for the hydrolysis of adenosine triphosphate (ATP) to adenosine diphosphate (ADP) and phosphate ($\Delta G^{o'}$ = -30.5 kJ/mol). What types of bonds are involved in each of these reactions? What can you conclude from these data?

Chapter 9 The Importance of Energy Changes and Electron Transfer in Metabolism

3. ATP is universally considered a carrier of chemical energy, which can be used in metabolic reactions; however, the cell also contains the other nucleoside triphosphates, guanosine triphosphate (GTP), thymidine triphosphate (TTP), cytidine triphosphate (CTP) and uridine triphosphate (UTP).
   a. Can these molecules also serve as carriers of chemical energy in a similar manner?
   b. Suggest a possible reason why they are not used as energy carriers in many cases?

4. Consider the reactions below and the associated free energy values:

		ΔG (kJ/mol)
i.	A + B ⇌ C	+ 12
ii.	C + D ⇌ E	− 32

   a. Will either of these reactions occur spontaneously?
   b. If reaction (i) were written in the opposite way (C ⇌ A + B), would the reaction be spontaneous? What would be the value of ΔG?
   c. Hess's Law states that the energy change in a chemical reaction is the same whether the reaction takes place in one step or several steps. This law can be used to theoretically calculate the free energy change for a reaction that has never been studied. Using Hess's Law, determine the free energy change for the following reaction.

   $$A + B + D \rightleftharpoons E$$

Chapter 9 The Importance of Energy Changes and Electron Transfer in Metabolism

5. The list below contains the ΔG^o values for reactions at 25°C.

ΔG^o (kJ/mol)
34.3
11.4
0.0
-5.7
-11.4
-34.3

   a. Calculate the corresponding K_{eq} values from each ΔG^o value in the list.
   b. An interaction between two molecules, which involves hydrogen bonding, contributes an energy of interaction (ΔG^o) of about 20 kJ/mol. How much greater would the equilibrium constant for an interaction be if an additional hydrogen bond were involved in the interaction?

6. If glucose is completely oxidized to CO_2 and H_2O in a laboratory experiment, it liberates 2867 kJ/mol of heat energy ($\Delta G^{o'}$ = -2867 kJ/mol).
   a. Write the balanced equation for this reaction.
   b. Is this an exergonic or endergonic reaction?
   c. What is the form of the energy that is produced?
   d. This same process is carried out enzymatically in human cells. Calculate the $\Delta G^{o'}$ for this cellular reaction? What form or forms does the liberated energy take within the cell?

7. In the process of glycolysis, a series of reactions occurs. One of the reactions involves the splitting of the 6-carbon modified sugar, fructose-1,6-bisphosphate, to produce two 3-carbon molecules, dihydroxyacetone and glyceraldehyde-3-phosphate. The standard free energy for this reaction is +23.9 kJ/mol; however, this reaction occurs spontaneously *in vivo*. Briefly explain.

8. Indicate the oxidant and the reductant in both the reactants and products for each of the reactions below. If there is no oxidation or reduction, indicate that no redox reaction occurs.

   a. ATP + H_2O ⇌ ADP + P_i
   b. Glyceraldehyde 3-phosphate + P_i + NAD^+ ⇌ 1,3-bisphosphoglycerate + NADH + H^+

Chapter 9 The Importance of Energy Changes and Electron Transfer in Metabolism

c. $6 H_2O + 6 CO_2 \rightleftharpoons C_6H_{12}O_6 + 6 O_2$
d. $Mg + CuCl_2 \rightleftharpoons MgCl_2 + Cu$
e. $2 H_2O_2 \rightleftharpoons 2 H_2O + O_2$

9. Match the terms in the left column with the appropriate word or phrase in the right column.

a. $FADH_2$ _____
b. ATP _____
c. Acetyl-CoA _____
d. Catabolic process _____
e. NAD^+ _____
f. CoA _____
g. Endergonic process _____
h. Oxidation _____
i. Anabolism _____
j. CoA, FAD^+, NADH _____

1. An exergonic, oxidative process
2. Exhibits a positive ΔG
3. Transfers 2 H^+ and 2 e^-
4. Contains both ester and acid anhydride bonds
5. Removes hydrogens
6. A high energy thioester
7. Has a reactive sulfhydryl group
8. An oxidizing agent
9. Has ADP in its structure
10. Biosynthesis

10. An endergonic reaction (a) having a ΔG° of 20 kJ/mol is coupled to an exergonic reaction (b) which has a ΔG° of -52 kJ/mol. Calculate the equilibrium constant at 25°C for each individual reaction and the coupled reaction.

11. The reaction in which glucose is converted to lactate and 2 H^+ has a ΔG° value of -196 kJ/mol. Consider the coupling of this reaction to the reaction for the production of ATP, in which 2 ATPs are produced. What is the standard free energy for the coupled reaction?

12. In the up-coming chapters, cellular metabolism will be presented in terms of the catabolic and anabolic reactions in the cells. The metabolic pathways will include glycolysis, the citric acid cycle, electron transport and oxidative phosphorylation, fatty acid synthesis and more. As alluded to in Chapter 1, catabolic processes produce ATP, the universal energy carrier, while anabolic processes utilize ATP in the biosynthesis of macromolecules. In this complex array of interrelated processes, each pathway is regulated in a number of ways. One of the more important regulatory signals in the cell is the level of ATP.

Chapter 9 The Importance of Energy Changes and Electron Transfer in Metabolism

An expression of the energy state of the cell is the **energy charge**. It is the mole fraction of the adenosine nucleotides in the cell, which have high-energy phosphates. The expression is:

Energy charge = {[ATP] + 1/2[ADP]}/ {[ATP] + [ADP] + [AMP]}

The ADP concentration is multiplied by 1/2 because it has only one half the number of high-energy bonds found in ATP. This ratio can theoretically vary from 1.0 (fully charged) to 0 (completely discharged). The analogy of the cellular energy charge to an automobile battery is often made. The energy charge in the cell, however, is held within a limited range, at about 0.9.

In order to regulate the energy charge in the cell, ATP and ADP are allosteric effectors of regulatory enzymes in a number of the pathways. One such regulatory enzyme in the catabolic glycolysis pathway is phosphofructokinase.

- a. What is the energy charge when the [ATP] = 4 mM, [ADP] = 1 mM and the [AMP] = 0.15 mM?
- b. What type of effectors are ATP and ADP on phosphofructokinase and what effect would each exert on the production of ATP and the energy charge?

13. Draw the structures for coenzyme A and the oxidized and reduced forms of the coenzymes, NAD+ and FAD. Indicate the reactive sites in each molecule.

ANSWERS TO EXERCISES

1.
   - a. True.
   - b. True.
   - c. True.
   - d. False. The free energy change (ΔG) is zero. The standard free energy is related to the equilibrium constant.
   - e. True.
   - f. True.
   - g. False. The ΔG value depends on the experimental conditions.
   - h. True.

Chapter 9 The Importance of Energy Changes and Electron Transfer in Metabolism

- i. True.
- j. False. The value of thermodynamic quantities such as ΔG, ΔH, and ΔS are unrelated to the rate of the reaction.
- k. False. This is not a spontaneous process and therefore the ΔS_{univ} value will be negative and the ΔG value will be positive.
- l. False. Hydrophobic interactions are driven primarily by the increase in entropy.
- m. True.
- n. True.

2. a. Use the relationship, $\Delta G' = \Delta G^{o'} + 2.303\, RT \log K_{eq}$
   Equilibrium is defined as $\underline{\Delta G' = 0}$.
   The equation reduces to:
   $$\Delta G^{o'} = -2.303 RT \log K_{eq}$$
   $$= -2.303(8.31\ \text{J mol}^{-1}\text{K}^{-1})(298\text{K})\ \log(300)$$

   $\underline{\Delta G^{o'} = -14.2\ \text{kJ/mol}}$

   b. The value of $\Delta G^{o'}$ for the hydrolysis of ATP is -30.5 kJ/mol. This is greater than twice the value for the hydrolysis of AMP. The difference in energy lies in the type of bond that is hydrolyzed. In the case of ATP, an acid anhydride bond is involved, while in AMP hydrolysis, an ester bond is broken. This is a general result of great significance in biochemistry because as will be seen in subsequent chapters, an anhydride bond is considered a "high energy" bond. The hydrolysis of an anhydride bond is used to generate metabolic energy, while an ester bond is not a "high energy" bond and is not a source of energy to the cell.

Note in the following figures, the type of bond hydrolyzed in ATP and AMP.

Chapter 9 The Importance of Energy Changes and Electron Transfer in Metabolism

Adenosine triphosphate (ATP) — Anhydride Bond

(AMP) Adenosine monophosphate — Phospho*ester* Bond

3. a. Any of the nucleoside triphosphates could be used as carriers of energy. In all cases, the hydrolysis of the terminal phosphate (an anhydride bond) would produce the same amount of energy as in the hydrolysis of ATP.
   b. ATP is primarily used because it is more abundant in the cell than the other nucleoside triphosphates.

4. a. Reaction (i) is not spontaneous since the free energy is positive. Reaction (ii), on the other hand, will take place spontaneously since the free energy change is negative.

Chapter 9 The Importance of Energy Changes and Electron Transfer in Metabolism

b. ΔG for any reverse reaction has the same ΔG value as that for the forward reaction, but with the opposite sign. The ΔG value for the reverse reaction is -12 kJ/mol, thus, the reaction would be spontaneous if written in this direction.

c. The free energy for the reaction, A + B + D = E, can be determined in this case by simply summing the ΔG values for reactions (i) and (ii). Therefore, the ΔG for this reaction would be -20 kJ/mol. This is an example of an endergonic reaction **coupled** to an exergonic reaction. As a consequence of the coupling, the overall free energy change for the process is negative and the reaction will occur spontaneously. This type of coupling process, especially when involving the hydrolysis of ATP, is very important in biochemistry. The hydrolysis of ATP is coupled to reactions as diverse as transporting ions across membranes and the process of muscle contraction.

5. a.

ΔG° (kJ/mol)	K_{eq}
-34.3	$1 \times 10^{+6}$
-11.4	$1 \times 10^{+2}$
0.0	0.0
5.7	1×10^{-1}
11.4	1×10^{-2}
34.3	1×10^{-6}

b. Insertion of the ΔG°' value of -20 kJ/mol for the hydrogen bonding interaction energy into the equation, ΔG°' = -2.303RT log K_{eq}, yields a K_{eq} value of about $3.1 \times 10^{+3}$. This K_{eq} value would correspond to **the increase** in the value of K_{eq} for an interaction involving an additional hydrogen bond.

6. a. $C_6H_{12}O_6 + 6 O_2 \longrightarrow 6 CO_2 + 6 H_2O$
b. This is an exergonic reaction - a spontaneous reaction with a negative free energy change.
c. The energy from exergonic reactions can be utilized to produce work. In the laboratory experiment, the liberated energy is in the form of heat. No useful work is accomplished.

Chapter 9 The Importance of Energy Changes and Electron Transfer in Metabolism

    d.   Since the same reaction is being considered, whether it occurs in a cell or in the laboratory, an identical amount of energy ($\Delta G^{o'}$) is liberated. This exemplifies a characteristic of a thermodynamic state function. That is, the value of a change in a state function (whether ΔG, ΔH, or ΔS) is independent of the route taken to accomplish the reaction (in this case, chemical versus enzymatic). However, all the energy liberated in the cellular reaction is not lost as heat. A considerable fraction of the energy can be utilized to assist some endergonic cellular reactions to occur. As will become evident in subsequent chapters, this energy is in the form of ATP.

7.   The standard free energy ($\Delta G^{o'}$) gives the free energy value at standard conditions (concentrations at 1 M for solutes, except for $[H^+]$, which is at 1×10^{-7} M). Such standard conditions are not relevant in the cell. The thermodynamic quantity of actual interest and importance is the change in free energy (ΔG). The concentrations in the cell are such that the free energy is negative and therefore, a spontaneous reaction occurs.

8.   a.   This is a hydrolysis reaction, not a redox reaction.
       b.   Glyceraldehyde 3-phosphate and NADH are reducing agents and therefore will be oxidized. NAD^+ and 1,3-bisphosphoglycerate are oxidizing agents and will be reduced. Any reaction which involves $NAD^+/NADH$ or $FAD/FADH_2$ are redox reactions.
       c. H_2O and $C_6H_{12}O_6$ (glucose) are oxidized, while CO_2 and O_2 are reduced.
       d.   Mg and Cu will be oxidized, while $CuCl_2$ and $MgCl_2$ are reduced.
       e.   In this reaction, the reactant, hydrogen peroxide, **disproportionates** to H_2O and O_2. That is, the reactant undergoes both a self-oxidation and reduction. Therefore, H_2O_2 and O_2 are reduced, while H_2O_2 and H_2O are oxidized.

9.

    a.   3   f.   7
    b.   4   g.   2
    c.   6   h.   5
    d.   1   i.   10
    e.   8   j.   9

Chapter 9 The Importance of Energy Changes and Electron Transfer in Metabolism

10. Reaction a. $\Delta G^o = 20$ kJ/mol.

$$\Delta G^o = -2.303\, RT \log K$$
$$20{,}000 \text{ J/mol} = -2.303\, (8.31 \text{ J/mol}^{-1}\, K^{-1})\, (298K) \log K$$
$$\log K = -3.5$$
$$\underline{K = 3 \times 10^{-4}}$$

Reaction b. $\Delta G^o = -52$ kJ/mol
$$\log K = 9.1$$
$$\underline{K = 1 \times 10^{9}}$$

According to Hess's Law, thermodynamic state functions, such as G, H and S for reactions, can be simply added to obtain the quantity for the coupled or resultant combined reaction.

Coupled reaction with $\Delta G^0{}_{total} = \Delta G^o{}_1 + \Delta G^o{}_2$
$$= 20 \text{ kJ/mol} + (-52 \text{ kJ/mol})$$
$$= -32 \text{ kJ/mol}$$

$$\log K = 5.61$$
$$\underline{K = 4 \times 10^{\pm 5}}$$

By coupling reactions (a) to (b), the equilibrium constant for the reaction increases by 9 orders of magnitude.

11. The individual reactions of interest are:
    Glucose $\rightleftharpoons$ 2 Lactate$^-$ + 2 H$^+$ $\Delta G^o{}_1 = -196$ kJ/mol
    ATP + H$_2$O $\rightleftharpoons$ ADP + P$_i$ $\Delta G^o{}_2 = +30$ kJ/mol

    The overall reaction is:
    Glucose + 2 ATP + 2 H$_2$O $\rightleftharpoons$ 2 Lactate$^-$ + 2 ADP + 2 P$_i$ + 2 H$^+$

    The overall free energy for the reaction is:
    $$\Delta G^o{}_{overall} = \Delta G^o{}_1 + 2\, \Delta G^o{}_2$$
    $$= -196 + 2(+30)$$
    $$= \underline{-136 \text{ kJ/mol}}$$

This shows that the reaction for glycolysis (glucose to lactate) is an exergonic reaction which can be used to produce 2 molecules of ATP and remain an exergonic reaction.

Chapter 9 The Importance of Energy Changes and Electron Transfer in Metabolism

12. a. Energy charge = (4 mM + 0.5 mM)/(4 mM +1 mM + 0.15 mM)
       Energy charge = 0.87
    b. The glycolysis pathway is associated with the oxidation of glucose and is catabolic in character. It is involved in the generation of ATP. When ATP levels are low and ADP levels are high, ADP interacts with phosphofructokinase to increase the activity of this enzyme and as a result to increase the level of glycolysis and ATP production. ADP is therefore a positive allosteric effector. The increased level of ATP will increase the energy charge. On the other hand, when ATP levels are high, ATP acts as an negative effector and interacts with phosphofructose to inhibit its activity. The allosteric effectors, ADP and ATP, are regulators of metabolism.

13. a.

Coenzyme A (CoA)

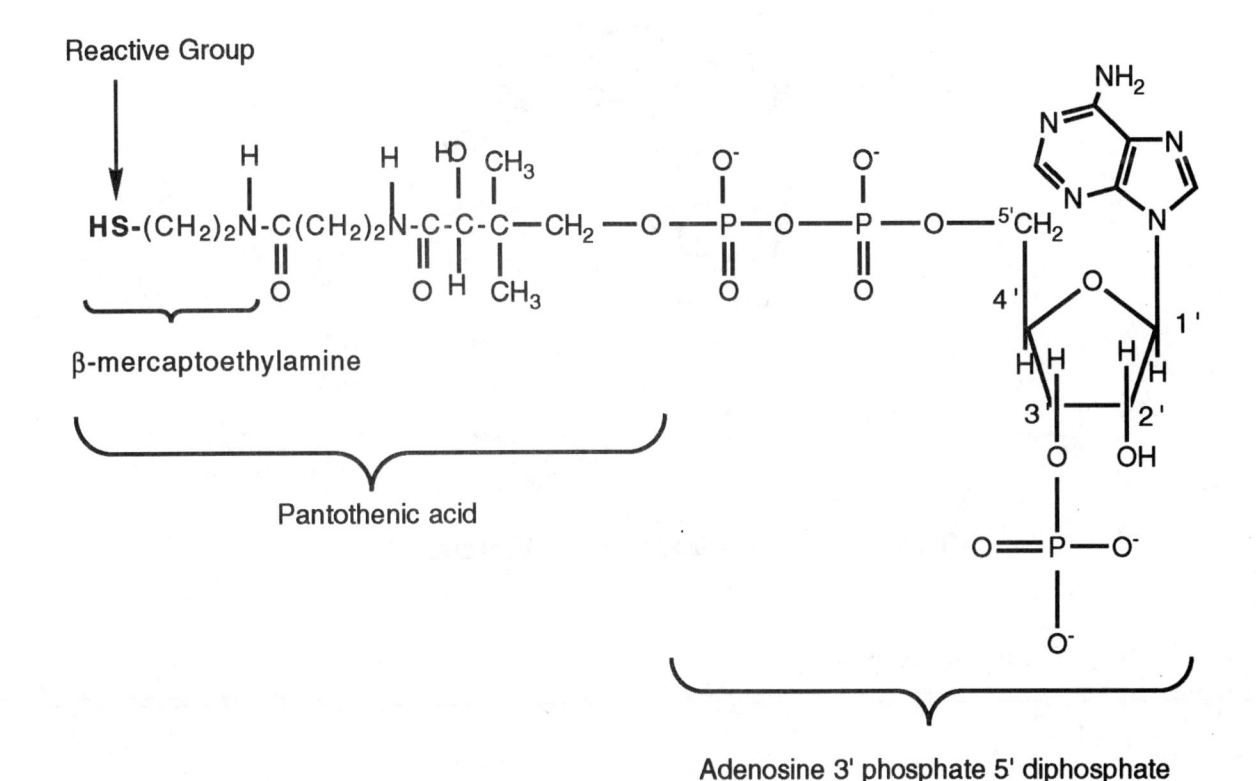

Reactive Group

HS-(CH$_2$)$_2$N-C(CH$_2$)$_2$N-C-C-C-CH$_2$—O—P—O—P—O—5'CH$_2$

β-mercaptoethylamine

Pantothenic acid

Adenosine 3' phosphate 5' diphosphate

Chapter 9 The Importance of Energy Changes and Electron Transfer in Metabolism

13b.

The reduced and oxidized forms - NADH and NAD⁺

c. See next page.

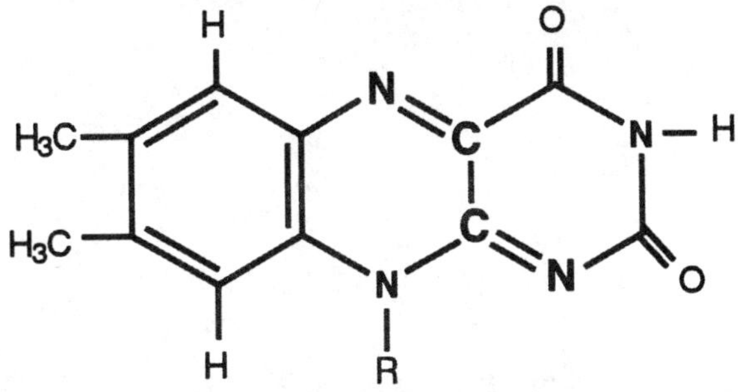

Chapter 9 The Importance of Energy Changes and Electron Transfer in Metabolism

Oxidized form of FAD

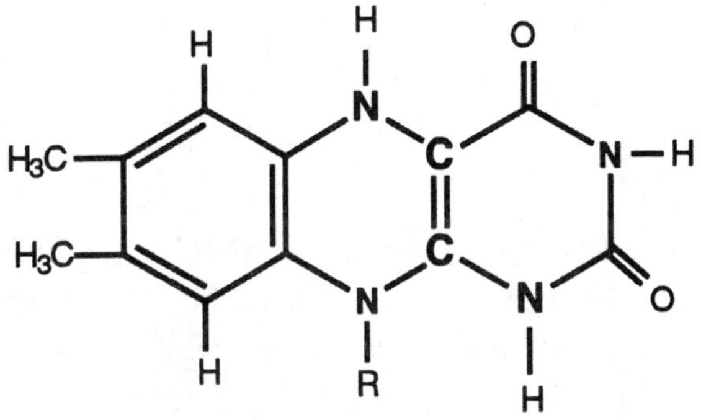

Reduced form of FADH$_2$

The reactive region is shown with the carbon and nitrogen atoms in the larger type.

10

Carbohydrates

Carbohydrates are polyhydroxy aldehydes or polyhydroxy ketones. Their primary function is to serve as an energy source in plants and animals, in addition to acting as essential structural elements in plants. The chemical reactivity of carbohydrates is associated with the main functionalities - those of the alcohol, aldehyde and/or ketone, hemiacetal and hemiketal - found in the carbohydrates. Carbohydrates can be portrayed in the **Fisher projection formulas**, or alternatively, in the **Haworth projection formulas** for the cyclic structures. Chiral centers abound in the aldoses and ketoses, with stereoisomers in the form of enantiomers, diastereomers, epimers and anomers. Linkage of two or more monosaccharides to form di-, oligo- or polysaccharides (glycosides) involves the formation of one or more glycosidic bonds, with the conversion of a terminal hemiacetal to an acetal linkage. These glycosidic backbones may be either α- or β- linkages, with the connectivity most often being between the C1-carbon on one monosaccharide and the C4-carbon of the other. Additional diversity in the structure and the function of these molecules is derived through (1) chemical modification of the monosaccharides, (2) many different monosaccharides in the make-up of a polysaccharide, together with (3) additional or different types of glycosidic linkages between the monomeric units. **Glycoproteins** play a critical role in eukaryotic organisms, which exhibit an immune response. Both the antibodies (i.e., the immunoglobulin molecules) and the antigenic determinants in many

Chapter 10 Carbohydrates

antigens are glycosylated, with the modified peptides serving to contribute to the specificity of the interaction.

CHAPTER OBJECTIVES

1. Be prepared to recognize stereoisomers and distinguish between enantiomers and diastereomers, including epimers and anomers.
2. Outline the characteristics of a hemiacetal, an acetal, a hemiketal, and a ketal.
3. Draw the Fisher projection formulas for aldoses and ketoses and classify the molecules according to the type of stereoisomer.
4. Convert Fisher projection formulas for aldohexoses into the corresponding Haworth formulas.
5. Point out in both the Fisher and Haworth formulas the carbon atom which defines the D- or the L-family of stereoisomers.
6. Indicate the chemical basis for a mono- or disaccharide being a reducing or nonreducing sugar.
7. Write the equations for monosaccharides undergoing (1) reduction, (2) esterification, and (3) glycoside formation.
8. Draw the structures for the reactants and products in the formation of sucrose, lactose, and maltose. Point out the anomeric carbon atoms and the type of linkages formed in the products.
9. Catalog the similarities and differences in the composition, structure and function of the major polysaccharides found in plants and animals.
10. Draw and name two of the major amino sugars and indicate in which polysaccharide they are found.
11. Characterize a typical glycoprotein and indicate some functions for them.

EXERCISES

1. Classify (e.g., as an aldotetrose) each of the following monosaccharides and indicate the number of chiral carbon atoms in each molecule.

Chapter 10 Carbohydrates

a.
```
    CHO
    |
  HO-C-H
    |
   H-C-OH
    |
  HO-C-H
    |
   CH₂OH
```

b.
```
    CHO
    |
   H-C-OH
    |
  HO-C-H
    |
  HO-C-H
    |
  HO-C-H
    |
  HO-C-H
    |
   CH₂OH
```

c.
```
   CH₂OH
    |
    C=O
    |
   CH₂OH
```

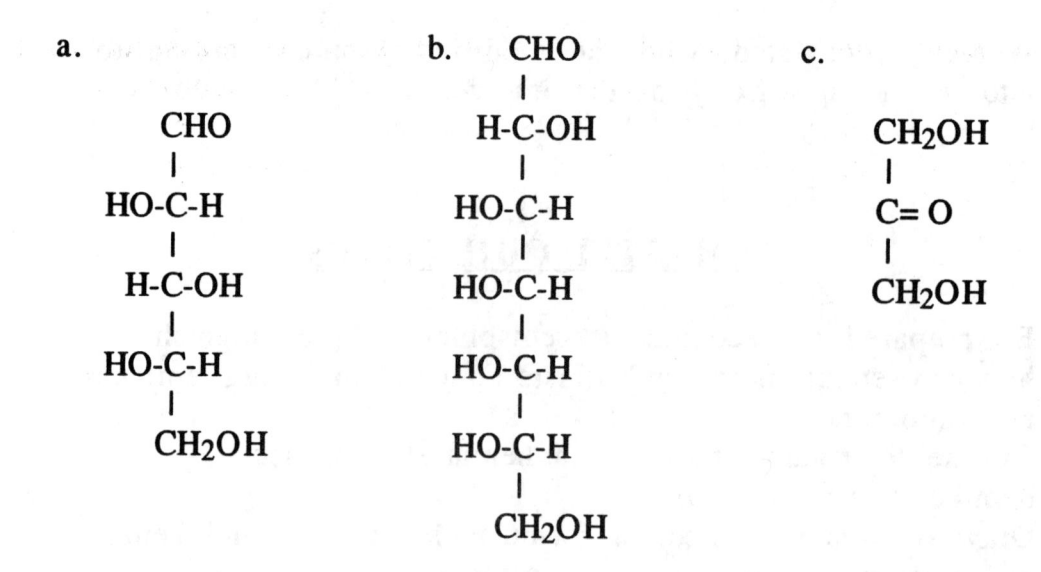

d.
```
    CHO
    |
   H-C-OH
    |
   CH₂OH
```

e.
```
        H
        |
        C=O
        |
   H—C—OH
        |
  HO—C—H
        |
   H—C—OH
        |
   H—C—OH
        |
       CH₂OH
```
D-Glucose

2. Indicate whether the monosaccharides in question 1 are in the D- or L-form, if either.

146

Chapter 10 Carbohydrates

3. Indicate which of the following molecules can be classified as a monosaccharide, disaccharide, aldehyde, ketone, hemiacetal, acetal, hemiketal or ketal or none of these. Some of the molecules can be classified in more than one category.

a.
```
    CH₃
    |
    C=O
    |
    CH₂
    |
    CH₃
```

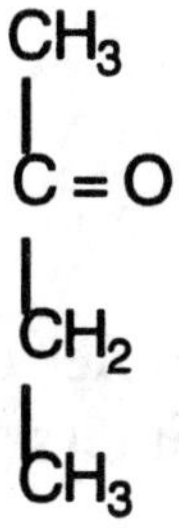

b. (tetrahydrofuran ring with O)

c. (benzene ring with –CHO)

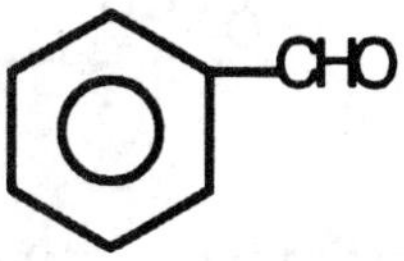

d.
```
      OCH₂CH₃
      |
 H₃C-C-H
      |
      OH
```

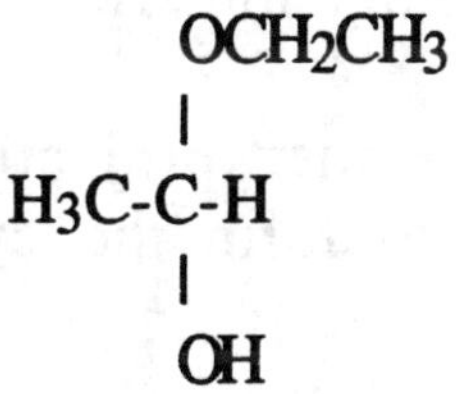

e.
```
      CHO
      |
 HO-C-H
      |
      CH₂OH
```

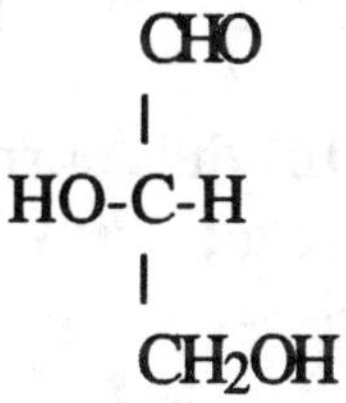

f.
```
      CH₂OH
      |
      C=O
      |
 HO-C-H
      |
  H-C-OH
      |
  H-C-OH
      |
      CH₂OH
```

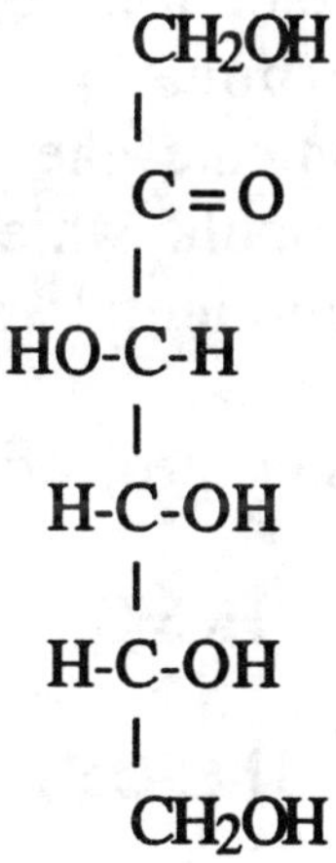

g.   Sucrose

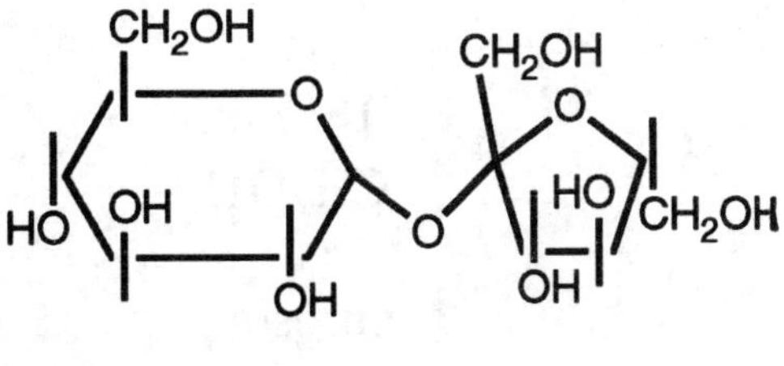

(structure of sucrose shown)

147

Chapter 10 Carbohydrates

h.   Lactose

$$\text{[structure of lactose: two pyranose rings joined by glycosidic bond, with CH}_2\text{OH, OH groups as shown]}$$

4. Define the terms diastereomer and epimer. Give an example of two diastereomers which are epimers and two diastereomers which are not epimers.

5. a. Draw the structure of α–D-glucose in the Haworth projection formula (ring) and point out the anomeric carbon.
    b. Indicate the atoms in the corresponding Fisher projection formula which will react with each other to form the cyclic hemiacetal structure.

6. The Fisher projections for D-ribose and L-gulose are shown below.

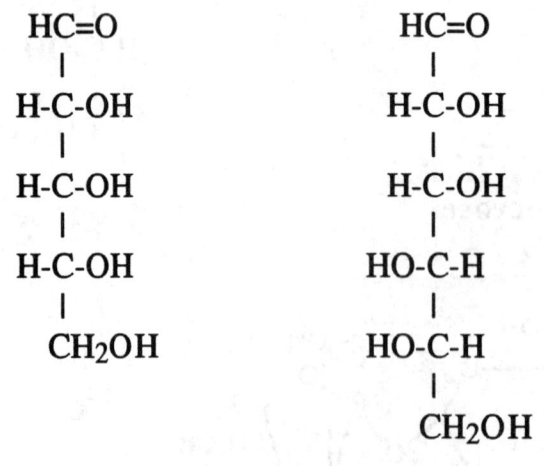

```
        HC=O              HC=O
         |                 |
        H-C-OH            H-C-OH
         |                 |
        H-C-OH            H-C-OH
         |                 |
        H-C-OH            HO-C-H
         |                 |
        CH₂OH             HO-C-H
                           |
                          CH₂OH

       D-ribose          L-gulose
```

a. Draw the Fisher projections for L-ribose and D-gulose.
b. Draw Haworth projections for D- and L-ribose in the α-form and for the D- and L-gulose in the β-form.
c. How many enantiomers are there for L-gulose?

Chapter 10 Carbohydrates

7. Which of the molecules shown in problem 3 will give a positive Tollen's test?

8. As discussed in the answer for question 7, a ketose gives a positive Tollen's test because it can be converted to an aldose through an enediol intermediate. Draw the structures for a ketose, an aldose, and an enediol and indicate which of these molecules are enantiomers or diastereomers.

9. Cellulose, the most abundant organic compound in the biosphere, is a linear, homopolysaccharide of D-glucose. Humans, however, can not digest it. Briefly explain this.

10. The antibiotic, penicillin, is prescribed to patients who have a bacterial infection. This drug binds covalently in the active site of glycoprotein peptidase. This enzyme is responsible for the cross-linking in the peptidoglycan chains, which is essential for the synthesis of the bacterial cell walls. Suggest how the action of penicillin kills bacteria.

11. Amylose is digested by a pair of enzymes.
    a. Name the hydrolytic enzymes and indicate the specific nature of the cleavage reaction and the products formed.
    b. Amylopectin requires (the action of) three enzymes for complete digestion. In addition to the two enzymes used to digest amylose, what is the additional enzyme and what is its role?

12. Human blood groups are classified as A, B, AB or O type.
    a. How are oligosaccharides important in this classification? Where are these glycoproteins found in blood cells?
    b. What monosaccharides are important in distinguishing the different blood types?

Chapter 10 Carbohydrates

13. Match the polysaccharide in the left column with the correct description on the right.

a. cellulose	_____	1.	polymer of primarily D-galacturonic acid
b. chitin	_____	2.	modified polysaccharide used as an anticoagulant
c. heparin	_____	3.	linear homopolysaccharide of D-glucose
d. pectin	_____	4.	main storage polymer for carbohydrates in animals
e. glycogen	_____	5.	found in the exoskeleton of insects

ANSWERS TO EXERCISES

1.   a. aldopentose, 3   d. aldotriose, 1
    b. aldoheptose, 5   e. aldohexose, 4
    c. ketotriose, 0

2. In the Fisher formula, a monosaccharide is a member of the **D-family** if the -OH group on the chiral carbon farthest from the C1 carbon is on the **RIGHT** side. For the **L-family,** the corresponding -OH group is on the **LEFT**.

    In the Haworth formula, the terminal -CH$_2$OH group is positioned **ABOVE** the ring in the **D-family** or alternatively **BELOW** the plane of the ring in the **L-family**. The key relationships between the groups on a Fisher and Haworth formula are:

    • Groups on the **RIGHT** in the Fisher formula are directed **DOWN** in the Haworth formula.
    • Groups on the **LEFT** in the Fisher formula are directed **UP** in the Haworth formula.

        a. L-form
        b. L-form
        c. no stereoisomers
        d. D-form
        e. D-form

Chapter 10 Carbohydrates

3.
   a. ketone
   b. none (this is an ether)
   c. aldehyde
   d. hemiacetal
   e. monosaccharide, aldehyde
   f. monosaccharide, ketone
   g. disaccharide, acetal
   h. disaccharide, acetal, hemiacetal

4. Diastereomers are any two stereoisomers which are not enantiomers. Epimers are diastereomers which differ in the configuration at only one of the chiral carbon atoms.

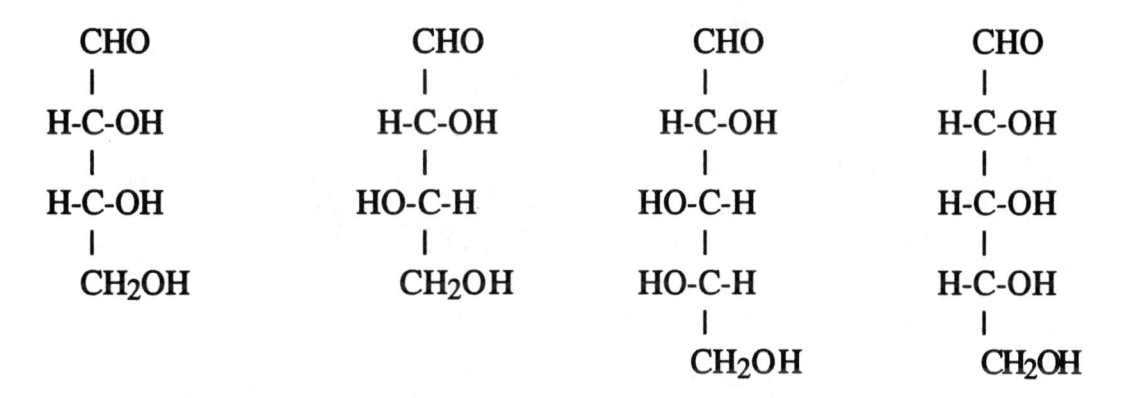

```
    CHO                CHO               CHO               CHO
     |                  |                 |                 |
   H-C-OH            H-C-OH            H-C-OH            H-C-OH
     |                  |                 |                 |
   H-C-OH            HO-C-H            HO-C-H            H-C-OH
     |                  |                 |                 |
    CH₂OH             CH₂OH            HO-C-H            H-C-OH
                                          |                 |
                                        CH₂OH             CH₂OH
```

Diastereomers which are epimers.   Diastereomers which are not epimers.

5. The Haworth projection formula for α–D-glucose is:

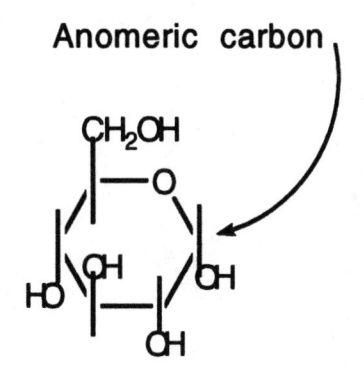

Anomeric carbon

(Haworth structure of α-D-glucose showing CH₂OH, ring O, OH groups, with anomeric carbon indicated)

151

Chapter 10 Carbohydrates

The Fisher projection for α-D-glucose is:

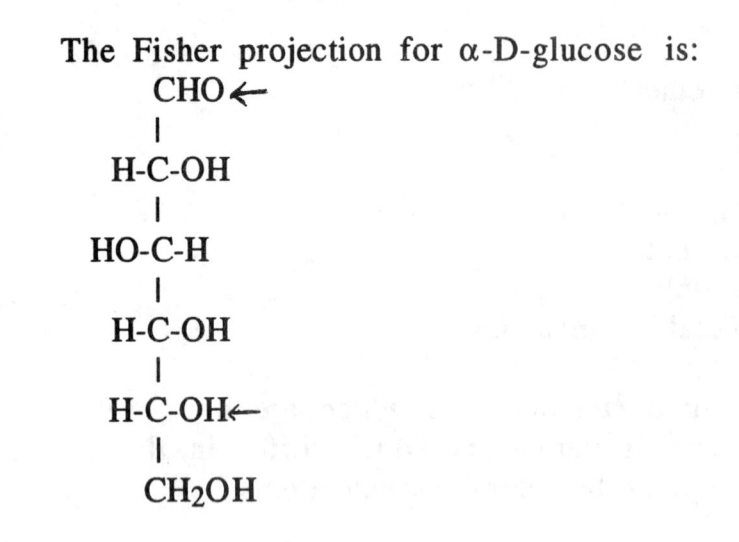

```
      CHO ←
       |
     H-C-OH
       |
     HO-C-H
       |
     H-C-OH
       |
     H-C-OH ←
       |
      CH₂OH
```

6. a.
```
      CHO                  CHO
       |                    |
     H-C-OH              H-C-OH
       |                    |
     H-C-OH              H-C-OH
       |                    |
     HO-C-H              HO-C-H
       |                    |
      CH₂OH              H-C-OH
                            |
                          CH₂OH

     L-ribose             D-gulose
```

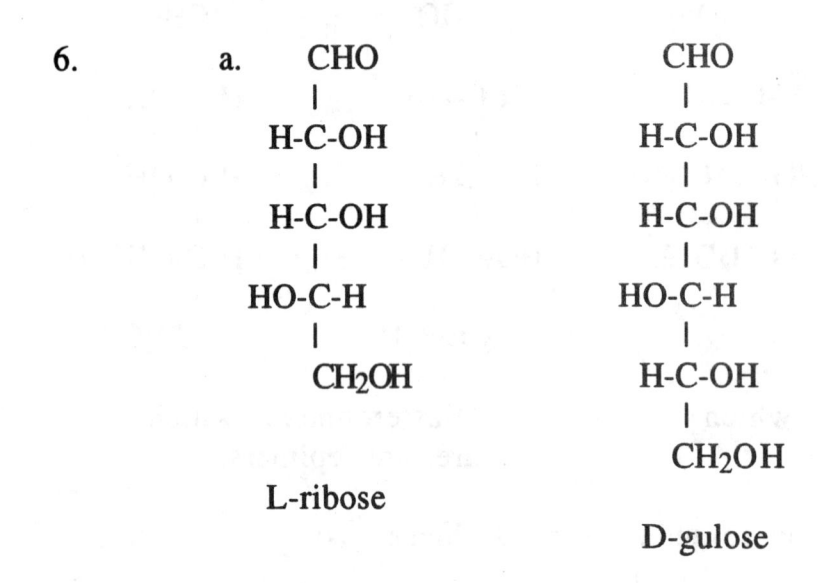

Chapter 10 Carbohydrates

b.

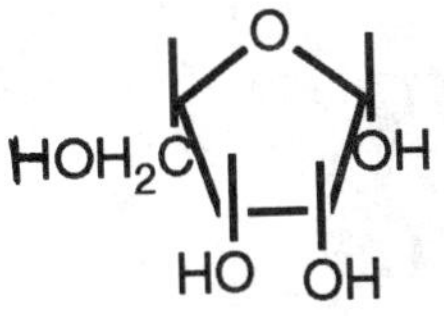

L-α-ribose

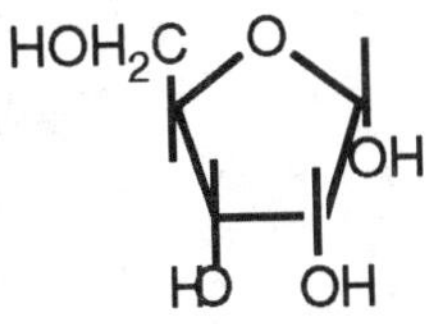

D-α-ribose

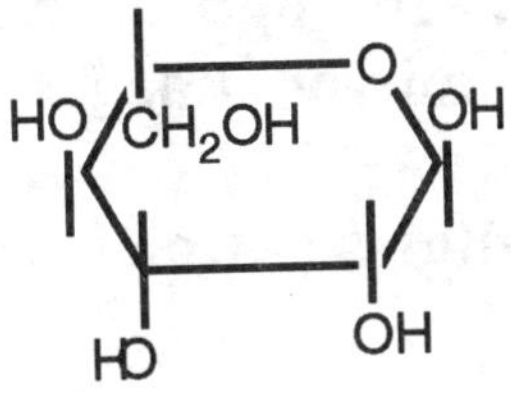

L-β-gulose

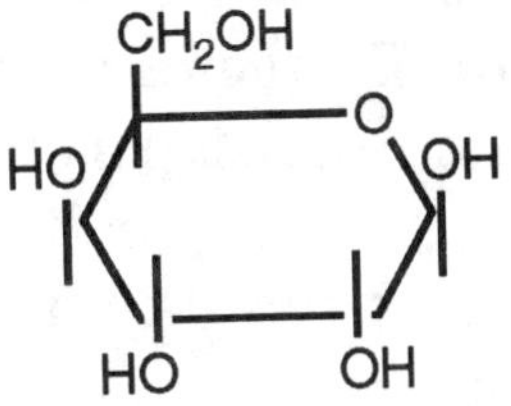

D-β-gulose

Enantiomers are PAIRS of stereoisomers which are nonsuperimposable mirror images. Therefore, there is only one enantiomer of L-gulose. It is D-gulose, which is shown above.

7. A positive Tollen's test indicates that the molecule contains an aldehyde which is readily oxidized to the corresponding acid.
   a. a ketone; it will not be oxidized.
   b. an ether; it will not be oxidized.
   c. an aldehyde; it will be oxidized.
   d. a non-cyclic hemiacetal; it will not be oxidized.
   e. an aldehyde; it will be oxidized.
   f. a ketohexose. This is an exception to the simple rule that ketones can not be oxidized. Ketones which have an -OH group on the adjacent carbon atom (i.e., an α–hydroxy ketone) are oxidized. Therefore, this monosaccharide, like all monosaccharides, whether an aldose or ketose, can be oxidized.
   g. contains only acetal functionalities which are not oxidized. This is an example of a non-reducing sugar.
   h. contains a terminal cyclic hemiacetal, which permits ring opening to yield an aldehyde. This is oxidized and is an example of a reducing disaccharide.

Chapter 10 Carbohydrates

8.

```
    CH₂OH              CHOH              CHO
     |                  ||                |
     C=O      ⇌       C-OH      ⇌      H-C-OH
     |                  |                 |
     R                  R                 R
   ketose            enediol            aldose
```

Note that there is both a rearrangement of electrons and hydrogens. These are diastereomers of D-fructose since they are optical isomers which are not enantiomers (i.e., nonsuperimposable mirror images).

9. Humans do not have the enzyme, cellulase, which can digest β(1-->4) glycosidic linkages.

10. Penicillin does not permit the complete synthesis of the bacterial cell walls. The peptidyl cross-links between the polysaccharide chains are essential for the structural integrity of the cell wall. Any segments of the cell wall without the cross-links are very susceptible to rupture by osmotic lysis; therefore, the bacteria will die.

11. Amylose is a linear polymer of glucose, having α(1-->4) glycosidic connecting bonds. The two enzymes involved in the digestion are:
    a. β-Amylase, an **exo**glycosidase, cleaves from the **non-reducing end** to produce **maltose.**
       α-Amylase, an **endo**glycosidase, hydrolyzes the glycosidic bond **within** the chain to produce **glucose and maltose.**
    b. Amylopectin is a branched polymer of glucose and therefore also has an α(1-->6) glycosidic bond at the branch points. The additional enzyme needed is the debranching enzyme, which hydrolyzes the α(1-->6) glycosidic linkages.

12. a. Glycoproteins are proteins which contain amino acid residues modified with carbohydrate residues. These glycoproteins reside on the surface of erythrocytes in the blood and determine the human blood group.

    b. L-Fucose is a component of the oligosaccharide in all the blood groups. N-acetylgalactosamine is found at the non-reducing

Chapter 10 Carbohydrates

end of the oligosaccharide in type A blood, while in type B, the N-Acetylgalactosamine is replaced by α-D-galactose. In type AB blood, both oligosaccharide forms exist, while in type O blood, neither form is found. Refer to Figure 10.26 in the text.

13. a. 3
    b. 5
    c. 2
    d. 1
    e. 4

11

Glycolysis

Glycolysis is the metabolic pathway in which one molecule of glucose is converted in a stepwise manner into two molecules of pyruvate. This anaerobic pathway represents one of the most primitive pathways and is the singularly most important exergonic pathway for the catabolism of glucose, not only in man, but also in plants and many microorganisms. In the process, energy is released, with part of it being converted into useful work in living organisms. Under anaerobic conditions, pyruvate can then be reduced to lactate in animals or microorganisms or alternatively to ethanol (alcoholic fermentation) in yeast. On the other hand, under aerobic conditions, pyruvate reacts with coenzyme A to produce CO_2 and acetyl-CoA, with the latter then shuttling into the citric acid cycle. Only in this latter case does glucose ultimately become completely oxidized to CO_2 and H_2O.

Chapter 11 Glycolysis

Some Key Figures Revisited

Figure 11.1 One molecule of glucose is converted to two molecules of pyruvate. Under aerobic conditions, pyruvate is oxidized to CO_2 and H_2O by the citric acid cycle and oxidative phosphorylation. Under anaerobic conditions, lactate is produced, especially in muscle. Alcoholic fermentation occurs in yeast. The NADH produced in the conversion of glucose to pyruvate is reoxidized to NAD^+ in the subsequent reactions of pyruvate.

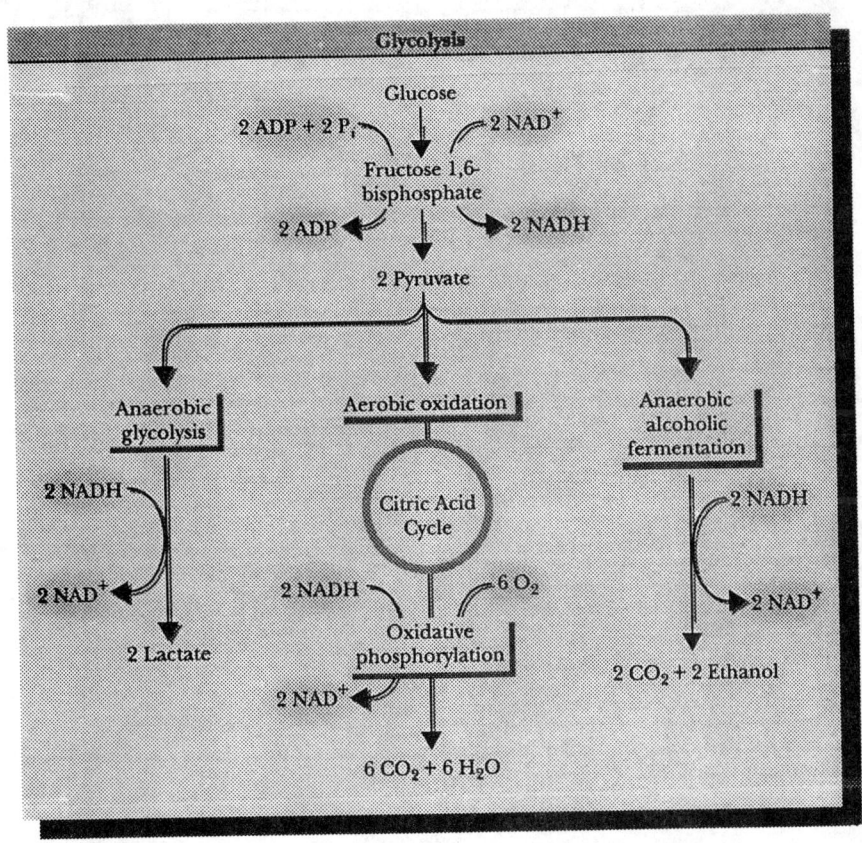

Chapter 11 Glycolysis

Figure 11.2 The pathway of glycolysis.

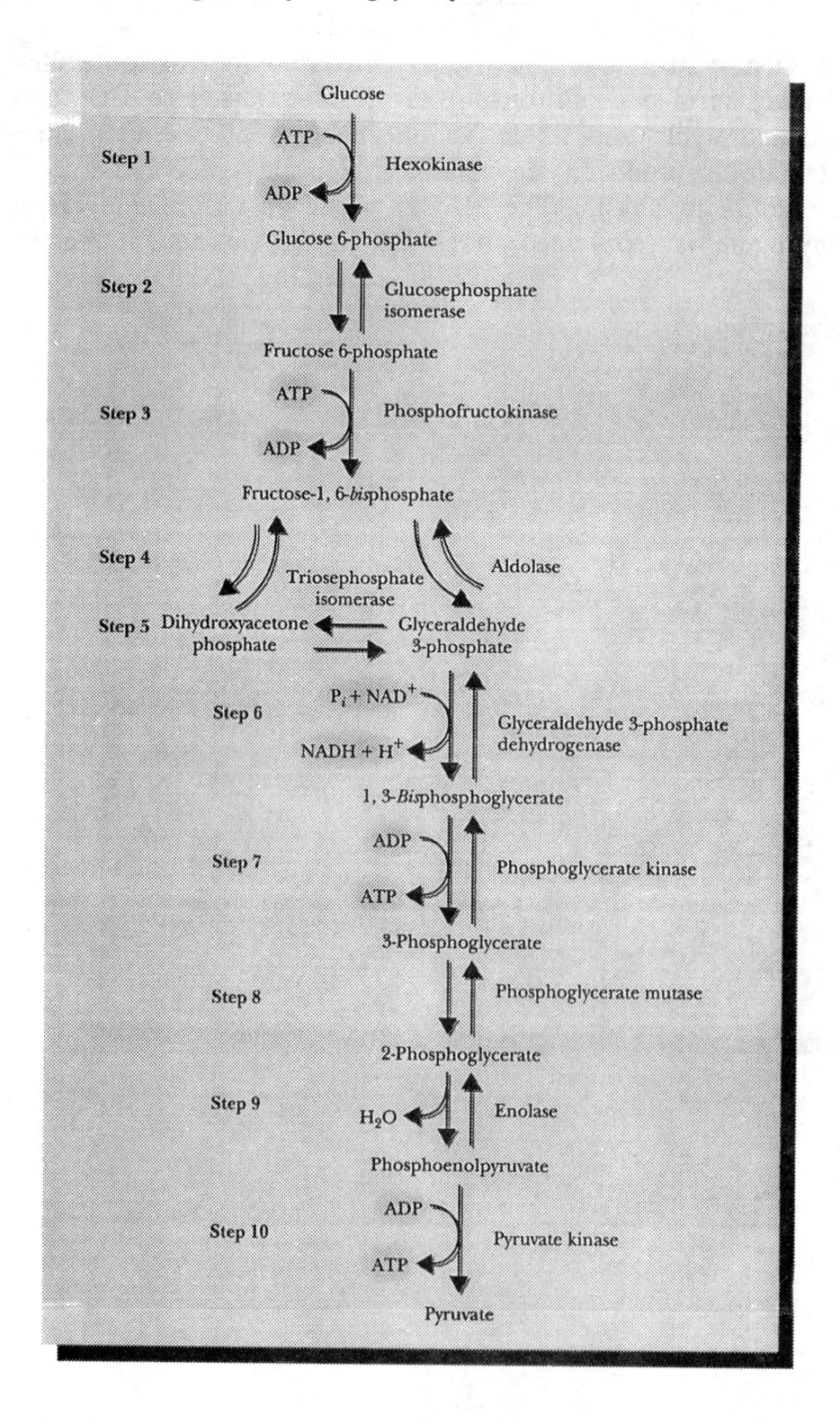

Glucose

Step 1: ATP → ADP, Hexokinase

Glucose 6-phosphate

Step 2: Glucosephosphate isomerase

Fructose 6-phosphate

Step 3: ATP → ADP, Phosphofructokinase

Fructose-1, 6-bisphosphate

Step 4: Aldolase

Step 5: Dihydroxyacetone phosphate ⇌ Glyceraldehyde 3-phosphate, Triosephosphate isomerase

Step 6: P_i + NAD^+ → NADH + H^+, Glyceraldehyde 3-phosphate dehydrogenase

1,3-Bisphosphoglycerate

Step 7: ADP → ATP, Phosphoglycerate kinase

3-Phosphoglycerate

Step 8: Phosphoglycerate mutase

2-Phosphoglycerate

Step 9: H_2O, Enolase

Phosphoenolpyruvate

Step 10: ADP → ATP, Pyruvate kinase

Pyruvate

Chapter 11 Glycolysis

CHAPTER OBJECTIVES

1. State the primary function of the glycolysis pathway.
2. Outline the steps involved in the conversion of glucose to pyruvate.
3. Name the types of reactions involved in glycolysis. Indicate which reaction involves the production or use of ATP, and state whether the reaction is exergonic or endergonic.
4. Name the enzyme that catalyzes each reaction in glycolysis and specify the enzymes which take part in the regulation of glycolysis.
5. Indicate the allosteric effector(s) for each regulatory enzyme and the effect that is produced by the interaction.
6. Write the reactions for an aldol condensation and Schiff base formation and present an example of each reaction as it occurs in glycolysis.
7. Specify the difference between substrate-level phosphorylation and phosphorylation involving ATP.
8. Outline the mechanism for the conversion of glyceraldehyde 3-phosphate to 1,3-bisphosphoglycerate by glyceraldehyde 3-phosphate dehydrogenase.
9. Specify the alternate metabolic fates of pyruvate and indicate the conditions under which they occur.
10. Indicate the extent to which the energy released by the oxidation of glucose in glycolysis is "captured." Into what molecular form is this energy captured?
11. Define the term isoenzyme; give a real example; and indicate how an isoenzyme discussed in this chapter is used in the diagnosis of a myocardial infarction.
12. Discuss how NAD^+ is regenerated in anaerobic glycolysis and why this is so important to the cell under anaerobic conditions.
12. Outline the mechanistic role of the coenzyme, thiamine pyrophosphate (TPP), in the enzymatic decarboxylation of pyruvate by pyruvate decarboxylase.
13. Compare the common and the unique features in the reductive pathways in which pyruvate is converted to lactate or ethanol.

Chapter 11 Glycolysis

EXERCISES

1. What are the final products of glucose metabolism in the following cells? In each case, write the equation showing the overall reaction.
   a. In actively metabolizing cells with a plentiful supply of O_2.
   b. In muscle cells of the leg after a strenuous 440 yard dash.
   c. In yeast cells under anaerobic conditions.

2. Consider the molecules and the reactions in the glycolytic pathway.
   a. How many carbon atoms are in the intermediate molecules?
   b. In a number of cases, the intermediate molecule is simply modified in a reaction. Considering this, what is the minimum number of "parent" molecules (C_n units) which are intermediates in the pathway?
   c. Comparing the values for the $\Delta G^{o'}$ in Table 11.1 (text) for the reactions in the glycolytic pathway, indicate the reactions which are highly exergonic?
   d. Of the ten reactions in glycolysis, how many actually involve an oxidation of an intermediate? Write these reactions.
   e. Oxidations in which **O_2 is directly** involved are catalyzed by enzymes classified as either **oxygenases** or **oxidases**. In cases in which oxidations occur in the absence of O_2, what is an alternate way of effecting an oxidation?
   f. It is found that regulatory steps in a pathway provide the bulk of the thermodynamic driving force which makes the overall pathway exergonic. Which reactions in Table 11.1 in the text are candidates for regulatory steps in glycolysis?
   g. Reactions labeled as 1, 7 and 10 in Table 11.1 are regulatory under physiological conditions in the cell, while reaction 3 is not, despite its large and negative $\Delta G^{o'}$ value. How can this be explained?

3. a. List the coenzymes involved in reactions discussed in this chapter?
   b. For each coenzyme, name at least one enzyme which requires it for activity.

Chapter 11 Glycolysis

    c. What is the essential function of each coenzyme?
    d. Draw the structure of each coenzyme and indicate the reactive site.

4. Indicate which reaction in the glycolytic pathway is catalyzed by the following enzymes:
    a. pyruvate kinase
    b. triosephosphate isomerase
    c. aldolase
    d. phosphoglyceromutase

5. What does it mean when it is stated that an enzyme catalyzes the "committed step" in the pathway?

6. Consider the conversion of glucose to lactate. Calculate the <u>average</u> oxidation state of the carbons in glucose and lactate, in order to determine the extent to which glucose has been oxidized at this point. Briefly explain the result.

ANSWERS TO EXERCISES

1. In all three cases, glucose will be converted via the 10 steps in the glycolytic pathway to pyruvate. After this point, the fate of the pyruvate will depend on the presence or absence of molecular O_2 and on the organism involved.
    a. In actively metabolizing cells, each pyruvate molecule loses one CO_2 and each remaining C2 unit (acetyl) is linked to coenzyme A to produce acetyl-CoA. Acetyl-CoA enters the citric acid cycle (Chapter 12) and becomes oxidized to two molecules of CO_2 (i.e., derived from the acetyl group). Upon completion of oxidative phosphorylation, the glucose has been completely oxidized to CO_2 and water.

$$C_6H_{12}O_6 + 6\ O_2 \longrightarrow 6\ CO_2 + 6\ H_2O \qquad \Delta G^{o'} = -2870\ \text{kJ/mol}$$

    b. After a strenuous run, the muscle cells of the leg will be deficient in O_2 and therefore, aerobic oxidation of pyruvate does not take place. In this environment, pyruvate will be **REDUCED** to lactate. Under these

Chapter 11 Glycolysis

conditions, known as anaerobic glycolysis, lactate will be the terminal product.

$C_6H_{12}O_6 \longrightarrow 2\ CH_3CHOHCOO^- + 2\ H^+ \qquad \Delta G^{o\prime} = -196$ kJ/mol
                         Lactate ion

    c.   In yeast cells, the pyruvate is decarboxylated to acetaldehyde (loss of CO_2) and the acetaldehyde is then **REDUCED** to ethanol (aldehyde reduced to an alcohol).

$C_6H_{12}O_6 \longrightarrow 2$ ethanol $+ 2\ CO_2 \qquad \Delta G^{o\prime} = -167$ kJ/mol
                    (C_2H_5OH)

    All three processes (reactions) are exergonic. However, there is incomplete oxidation of of glucose in the anaerobic conditions. In these conditions, the amount of energy liberated is less than 7% of that for the aerobic condition, in which the glucose is completely oxidized.

2.   a.   There are intermediate molecules that have either 6- or 3-carbon atoms.
    b.   For C6, the molecules are glucose and fructose. For C3, the molecules are dihydroxyacetone, glyceraldehyde, glycerate and pyruvate.
    c.   There are only 4 reactions that are highly exergonic. They are:

	REACTION	$\Delta G^{o\prime}$ (kJ/mol)
i.	glucose + ATP ---> glucose 6-phosphate + ADP	-16.7
ii.	fructose 6-phosphate + ATP --> fructose 1,6-bisphosphate + ADP	-13.8
iii.	1,3-bisphosphoglycerate + ADP ---> 3-phosphoglycerate + ATP	-37.6
iv.	phosphoenolpyruvate + ADP ---> pyruvate + ATP	-62.8

    d.   Of the 10 reactions, there is only one oxidation reaction. The reaction is:

glyceraldehyde 3-phosphate + NAD^+ + P_i ----->
                          1,3-bisphosphoglycerate + NADH + H^+

Chapter 11 Glycolysis

Since NAD⁺ is reduced to NADH, the other reactant must be oxidized.

- e. The oxidation in question 2d involves the transfer of a hydrogen from the substrate to NAD⁺. Alternate descriptions for an oxidation include the loss of a hydrogen atom or the loss of electrons.
- f. Steps 1, 3, 7 and 10 in Table 11.1 (refer to answer 2c also).
- g. Standard free energy values ($\Delta G^{o'}$) often differ significantly from those under physiological conditions ($\Delta G'$). Calculations show that under physiological conditions, the $\Delta G'$ value for reaction 3 is negative, but very small, while only reactions 1, 7 and 10 exhibit large and negative values. These three reactions regulate glycolysis.

3. a, b, c.

Coenzyme	Enzyme(s)	Function
NAD⁺/ NADH	Many dehydrogenases (ex., Glyceraldehyde 3-phosphate dehydrogenase; Lactate dehydrogenase)	NAD⁺ acts as an oxidizing agent (e⁻ acceptor) NADH acts as a reducing agent (e⁻ donor)
TPP	Pyruvate decarboxylase	Cleavage or formation of a carbon-carbon bond adjacent to a carbonyl carbon

Chapter 11 Glycolysis

d.

The reduced and oxidized forms of NADH and NAD⁺

Reduced form — Reactive site — Oxidized form

Thiamine pyrophosphate (TPP)

Chapter 11 Glycolysis

4. a. Pyruvate kinase catalyzes the conversion of phosphoenolpyruvate to pyruvate, with the production of ATP.
   b. Triosephosphate isomerase catalyzes the isomerization of the triosephosphates, dihydroxyacetone phosphate to glyceraldehyde-3-phosphate.
   c. Aldolase catalyzes the non-hydrolytic cleavage of the C_6, fructose 1,6-bisphosphate to two C_3 monosaccharides (trioses), glyceraldehyde-3-phosphate and dihydroxyacetone.
   d. Phosphoglyceromutase catalyzes the transfer of the phosphate in 3-phosphoglycerate from carbon-3 to carbon-2 to produce 2-phosphoglycerate.

5. A metabolic pathway is a step-wise set of reactions, which converts the initial substrate into the final product. There are often many points along the pathway at which an intermediate may be diverted from the major pathway and used for another purpose. This alternate reaction can occur only prior to the committed step in the pathway. The committed step is the reaction in which the reaction product completes the pathway and is used for the formation of the final product.

6. The molecular formula for glucose is $C_6H_{12}O_6$ and that for lactate (lactic acid) is $C_3H_6O_3$. Assume that the oxidation states for oxygen and hydrogen are fixed at -2 and +1, respectively; therefore, since glucose and lactic acid are neutral molecules, the average oxidation state for carbon in each molecule can be readily calculated.

Glucose: $6x + 12(+1) + 6(-2) = 0$ $x =$ oxidation state for carbon
           $x = 0$ Oxidation state for carbon in glucose is zero.

Lactate: $3x + 6(+1) + 3(-2) = 0$
           $x = 0$ Oxidation state for carbon in lactic acid is zero.

This calculation indicates that the average oxidation state for the carbons has not changed in going from glucose to lactic acid. Note that there is one oxidation step in glycolysis (glucose to pyruvate) and a subsequent reduction step from pyruvate to lactate. Even though there is a net oxidation from glucose to pyruvate, this is lost in the reduction of pyruvate to lactate.

12

The Citric Acid Cycle

The **citric acid cycle** serves as the hub of cellular metabolic activity. It provides the central oxidative pathway by which metabolic fuel molecules - the carbohydrates, amino acids and fatty acids - are catabolized and energy, in the form of ATP, is generated under aerobic conditions. In addition, the metabolic intermediates serve as precursors for the biosynthesis of such diverse molecules as proteins, lipids and the heme group. Acetyl-CoA, the intermediate formed by the breakdown of the fuel molecules, serves as the dominant "entry molecule" into the cycle. As a result of the eight reactions in the cycle, the acetate group is oxidized to two molecules of CO_2 and three molecules of NADH, and one molecule of $FADH_2$ and one molecule of GTP are produced. The NADH and $FADH_2$ molecules are subsequently oxidized in the electron transport chain by O_2 to generate additional ATP molecules. Regulation within the cycle occurs at three points involving the enzymes, citrate synthase, isocitrate dehydrogenase, and α-ketoglutarate. In addition, the activity of pyruvate dehydrogenase is under stringent control. Each of the four reactions associated with these enzymes exhibits a large negative $\Delta G^{o'}$ value, which is a general

Chapter 12 The Citric Acid Cycle

characteristic of a regulatory step. All of the reactions associated with the citric acid cycle take place in the matrix of the mitochondria. The **glyoxylate cycle**, found only in plants and some microorganisms, is a variant of the citric acid cycle and provides a unique route to carry out the net synthesis of carbohydrates from fat.

Some Key Figures Revisited

Figure 12.5 (next page) A recapitulation of the citric acid cycle. Note the names of the enzymes. The loss of CO_2 is indicated, as is the phosphorylation of GDP to GTP. The production of NADH and $FADH_2$ is also indicated.

Figure 12.7 The glyoxylate cycle. This pathway results in the net conversion of two acetyl-CoA to oxaloacetate.

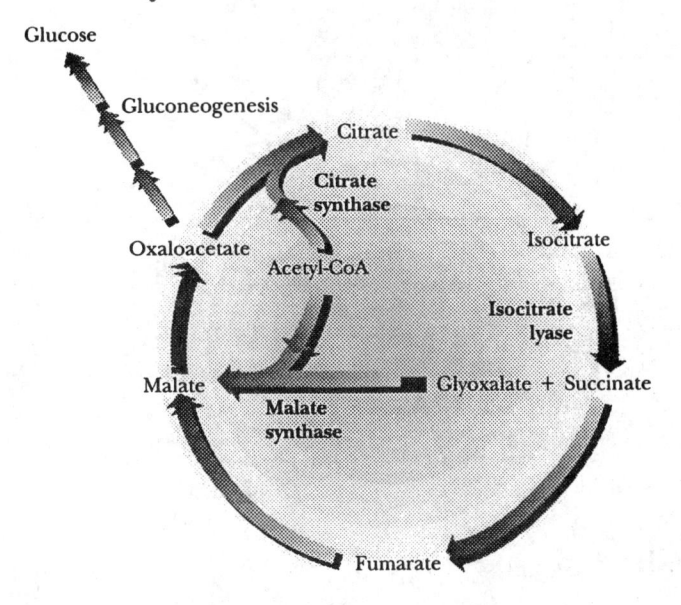

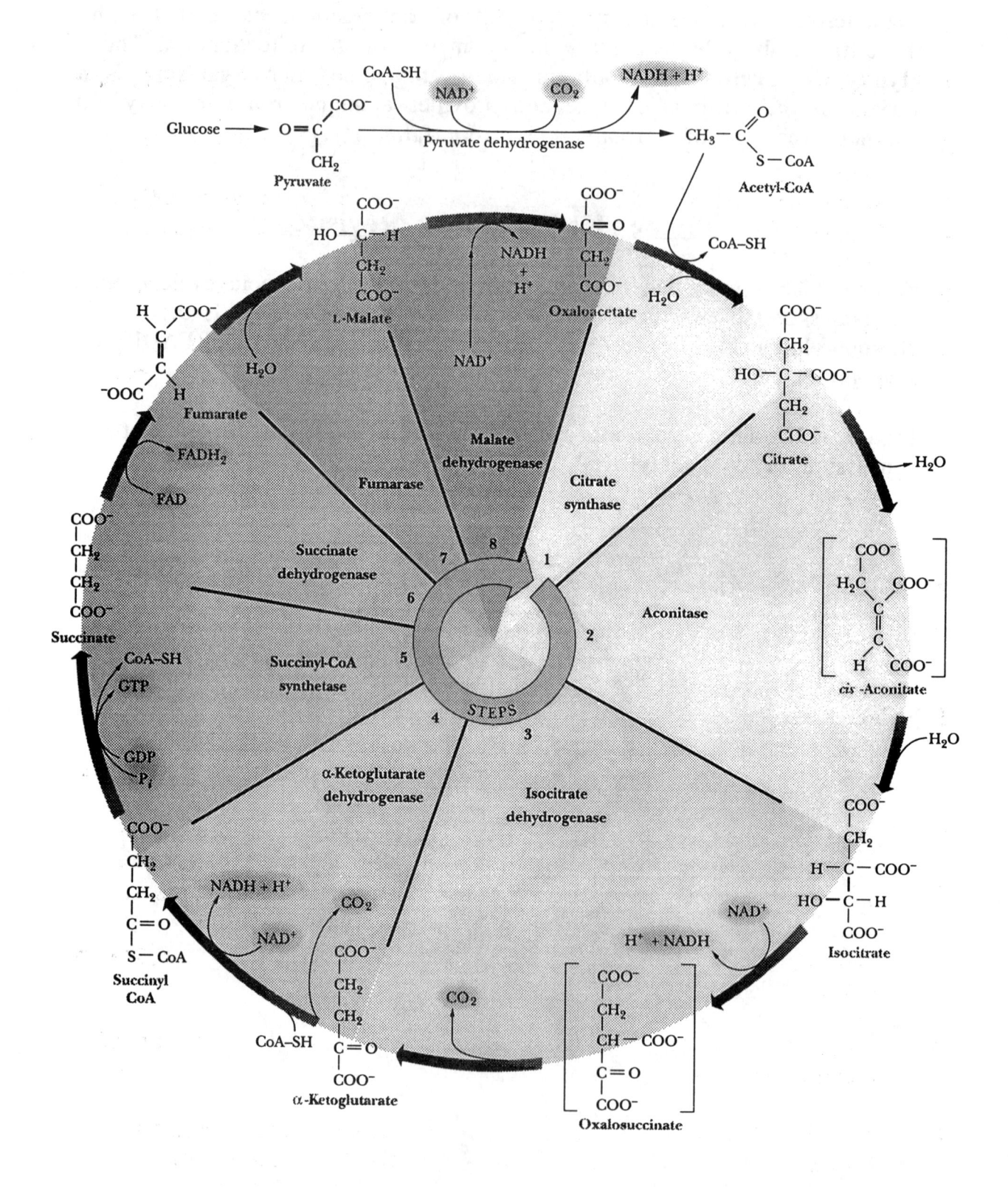

Chapter 12 The Citric Acid Cycle

168

Chapter 12 The Citric Acid Cycle

CHAPTER OBJECTIVES

1. State the primary catabolic and anabolic roles of the citric acid cycle and indicate the cellular location of the process.
2. Write the individual enzymatic reactions of the citric acid cycle, indicating which reactions are exergonic.
3. Write the overall reaction for the conversion of pyruvate to CO_2.
4. Name and draw the structures of the intermediates for the citric acid cycle.
5. List the enzymes associated with each reaction in the citric acid cycle and indicate the type of reactions that take place.
6. Indicate the regulatory enzymes in the cycle and the small molecules that act as inhibitors and activators for these allosteric enzymes.
7. List the components of the multienzyme complex, pyruvate dehydrogenase.
8. For the conversion of pyruvate to acetyl-CoA, write the five stepwise reactions involved and specify the enzymes and the coenzymes associated with each intermediate conversion.
9. Write the reactions involved in the glyoxylate cycle and state the importance of this cycle in plant cells.

EXERCISES

1. Name the di- and tricarboxylic acids that are in the citric acid cycle.

2. Pyruvate dehydrogenase and α-ketoglutarate dehydrogenase are both multienzyme complexes, which require multiple coenzymes and a number of reactions to catalyze the conversion of pyruvate and α-ketoglutarate, respectively, to their products. The mechanisms for these enzymic reactions are very similar. Suggest a reason for this observation.

3. Name the enzymes in the citric acid cycle and indicate the type of reaction that each catalyzes.

4. Acetyl-CoA is an example of a thioester. It has a $\Delta G^{o'}$ value for hydrolysis of -31.5 kJ/mol, while a typical $\Delta G^{o'}$ value for the hydrolysis of an ordinary ester is about -15 to -20 kJ/mol.
   a. Draw general structures for a simple thioester and an ester.

Chapter 12 The Citric Acid Cycle

    b.   Write the reaction for the hydrolysis of each type of ester.
    c.   Explain why the standard free energy of hydrolysis is so much greater for the thioester than for an ordinary ester.

5. In Table 12.1 (text), the reaction for the conversion of malate to oxaloacetate has a $\Delta G^{o'}$ value of +29.2 kJ/mol, indicating that this reaction is not spontaneous and has no thermodynamic driving force. However, the reaction takes place in the cell. Propose an explanation for this apparent anomaly.

6. In the reaction in which succinyl-CoA is converted to succinate, one molecule of GTP is produced. It is stated that the production of GTP is equivalent to that for ATP. How can you explain this statement in light of the following exchange reaction?

$$GTP + ADP \rightleftharpoons GDP + ATP$$

7. Indicate whether the following statements are true or false. If false, correct the statement.
    a.   Pyruvate dehydrogenase catalyzes the reversible conversion of pyruvate to acetyl-CoA.
    b.   The reactions in the citric acid cycle take place in the matrix of the mitochondria.
    c.   Pyruvate must be transported into the mitochondria before it can be acted upon by pyruvate dehydrogenase.
    d.   Lipoic acid is covalently attached to arginine by an amide linkage in the multienzyme pyruvate dehydrogenase complex.
    e.   Lipoic acid can act as both an oxidizing agent and an acyl group acceptor.
    f.   The initial reaction in the citric acid cycle and the glyoxylate cycle involves the formation of a carbon-carbon bond.
    g.   Thiamine pyrophosphate (TPP) is a coenzyme involved in the two oxidative decarboxylation reactions in the citric acid cycle.
    h.   The conversion of citric acid to isocitrate is necessary because citric acid cannot be oxidized, while isocitrate is readily oxidized.
    i.   After the completion of the citric acid cycle, in which glucose has been converted to 6 molecules of CO_2, virtually all of the energy and ATP have been produced.

Chapter 12 The Citric Acid Cycle

8. a. Draw a general diagram for the citric acid cycle and the glyoxylate cycle and show the intermediates. Explain briefly the important and the different features of each cycle.
   b. Where does the glyoxylate cycle take place in plant cells?
   c. Why can fat be metabolized to produce **net** carbohydrate synthesis in the glyoxylate cycle, but not in the citric acid cycle?
   d. Write the net equation for the reaction taking place in the glyoxylate cycle.
   e. Why is acetyl-CoA simply not converted to pyruvate to produce a pathway for the net synthesis of carbohydrate?
   f. Name and draw the structures for any molecules that are unique to the glyoxylate cycle.

9. a. What are the enzymes within the citric acid cycle which regulate the cycle according to metabolic needs?
   b. When a cell is in a metabolic state in which the ATP and the NADH levels are high, what can be said about the metabolic activity associated with the citric acid cycle?

10. What is the relationship between the metabolic activity of the cell and the energy charge of the cell, (characterized by the ATP/ADP ratio) and the NADH/NAD+ ratio in the cell?

ANSWERS TO EXERCISES

1. Tricarboxylic acids: Citric acid
                                    Isocitric acid
   Dicarboxylic acids: α-Ketoglutaric acid
                                    Succinate
                                    Fumarate
                                    Malate
                                    Oxaloacetate

2. Both reactions are similar in that they involve an oxidative decarboxylation of an α-keto acid to an acyl thioester.
   a.   pyruvate   ---->   acetyl-CoA

$$H_3C-\overset{O}{\underset{\|}{C}}-COO^- + CoA-SH \longrightarrow H_3C-\overset{O}{\underset{\|}{C}}-S-CoA + CO_2$$

Chapter 12 The Citric Acid Cycle

b. α-ketoglutarate ----> succinyl-CoA

$$^-OOC\text{-}CH_2\text{-}CH_2\text{-}\underset{\underset{O}{\|}}{C}\text{-}COO^- + CoA\text{-}SH \longrightarrow {}^-OOC\text{-}CH_2\text{-}CH_2\text{-}\underset{\underset{O}{\|}}{C}\text{-}S\text{-}CoA + CO_2$$

3.

	Enzymes	**Reaction Type**
1.	Citrate synthase	Condensation
2.	Aconitase	Isomerization (Note that this results from a hydration reaction followed by a dehydration reaction).
3.	Isocitrate Dehydrogenase	Oxidative Decarboxylation
4.	α-Ketoglutarate Dehydrogenase	Oxidative Decarboxylation
5.	Succinate Thiokinase	Hydrolysis and Phosphorylation
6.	Succinate Dehydrogenase	Dehydrogenation (i.e., oxidation)
7.	Fumarase	Hydration
8.	Malate Dehydrogenase	Dehydrogenation (i.e., oxidation)

(The numbers refer to the reactions in the Figure on page 176 for the citric acid cycle in the answer to question 8).

4. a. The structures for a simple thioester and an ordinary ester are shown below.

$$\underset{\text{Thioester}}{R\text{-}\underset{\underset{O}{\|}}{C}\text{-}S\text{-}R'} \qquad\qquad \underset{\text{Ester}}{R\text{-}\underset{\underset{O}{\|}}{C}\text{-}O\text{-}R'}$$

b.

$$R\underset{\underset{O}{\|}}{C}\text{-}SR' + H_2O \longrightarrow R\underset{\underset{O}{\|}}{C}\text{-}O^- + R'\text{-}SH$$

$$R\underset{\underset{O}{\|}}{C}\text{-}OR' + H_2O \longrightarrow R\underset{\underset{O}{\|}}{C}\text{-}O^- + R'\text{-}OH$$

Chapter 12 The Citric Acid Cycle

c. The $\Delta G^{o'}$ for the reaction is the difference between the free energy of the products and the reactants. This $\Delta G^{o'}$ is primarily a result of the difference in the resonance stabilization of the ester compared to the thioester. Because of the comparable sizes of the carbon and oxygen atoms (and the p orbitals in these atoms), there is a resonance stabilization in the ester; this is produced by the favorable overlap in the resulting π-orbital or electron overlap in the ester. Because the sulfur atom is larger than the carbon atom, there is little or no stabilization energy in the thioester. The resonance forms are shown below.

$$\begin{array}{cccc}
& \overset{\delta-}{O} & O & O \\
\overset{\|}{} & \overset{|}{}{\scriptstyle\delta+} & \overset{\|}{} & \overset{|}{} \\
R-C-OR' \longleftrightarrow R-C-OR' & & R-C-SR' \xleftarrow{X} R-C-SR'
\end{array}$$

   Resonance between the   Little or no resonance
     two forms   between the two forms

The stabilization energy lowers the energy of the ester relative to the thioester. The product molecules in both cases have comparable energies. The carboxylate anions are, in fact, the same; therefore, the $\Delta G^{o'}$ between the ester and carboxylate anion is smaller than the $\Delta G^{o'}$ for the thioester and carboxylate anion. The energy of hydrolysis is smaller for the ester than the thioester.

The following figure shows the proposed differential overlap in a (CO) and a (CS) bond.

Generalized picture of π-orbital overlap

    Ester   **Thioester**

Chapter 12 The Citric Acid Cycle

5. The positive value for the standard free energy change for this reaction indicates that it is clearly very unfavorable. The reaction within the cell is not at standard conditions. This factor, in addition to the influence of the subsequent reaction in the cycle, must be considered. The reaction immediately following this is for the condensation of acetyl-CoA and oxaloacetate, which is highly exergonic. This reaction depletes the concentration of oxaloacetate continuously during active metabolism and in this way, helps to drive the reaction to the right.

6. The production of GTP is equivalent to that for ATP because the exchange reaction has a $\Delta G^{o'}$ of zero. This indicates that there is effectively free exchange between the nucleoside diphosphate and nucleoside triphosphate (NDP and NTP) species.

7.
   a. False. This reaction is essentially irreversible.
   b. True.
   c. True.
   d. False. Lipoic acid is attached to a lysine residue.
   e. True.
   f. True.
   g. False. In the citric acid cycle, only the reaction for the conversion of α-ketoglutarate to succinyl-CoA uses TPP as a coenzyme. The oxidative decarboxylation of isocitrate does not. The conversion of pyruvate to acetyl-CoA also uses TPP, but this reaction is not in the citric acid cycle. Note, however, that the two reactions which utilize TPP are associated with the oxidative decarboxylation of α-keto acids (pyruvate and α-ketoglutarate).
   h. True. Recall that citric acid is a tertiary alcohol and does not undergo oxidation, while isocitrate is a secondary alcohol and readily undergoes oxidation. Refer to the summary of organic reactions in Chapter 2 in this book.
   i. False. Most of the molecules of ATP generated from the oxidation of glucose will be derived indirectly from the reoxidation of the NADH and $FADH_2$ molecules in the electron transport chain and oxidative phosphorylation.

Chapter 12 The Citric Acid Cycle

8.

a. In plants, fatty acids are produced as a result of the breakdown of lipids. The fatty acids are further catabolized to two carbon (acetyl) units and the acetyl-CoA is funnelled into the glyoxylate cycle. As shown in the figure below, reactions 1 and 2, and the enzymes that catalyze these reactions, are the same as in the citric acid cycle. Isocitrate is the product after these first two reactions. However, **reactions 3 to 7 are not part of the glyoxylate cycle**. Therefore, reactions 3 and 4, which eliminate two molecules of CO_2, are by-passed in the glyoxylate cycle. Isocitrate undergoes a cleavage reaction (**A**), catalyzed by isocitrate lyase, to produce glyoxylate and succinate. The glyoxylate undergoes a condensation reaction (**B**) with another molecule of acetyl-CoA to produce malate. This reaction is catalyzed by malate synthase. The malate then undergoes dehydrogenation (8) to produce oxaloacetate, which is then ready to initiate another turn of the citric acid or glyoxylate cycle. Note that there are only **five reactions in the glyoxylate cycle** (reactions 1, 2, **A**, **B** and 8).

The succinate which was produced in the glyoxysomes is transported to the mitochondria and there enters the citric acid cycle, where reactions 6 to 8 are carried out to complete the cycle. Since there is now an additional succinate molecule (four carbon unit) introduced into the citric acid cycle, once it is converted to oxaloacetate, it can be readily used in carbohydrate synthesis via gluconeogenesis.

Chapter 12 The Citric Acid Cycle

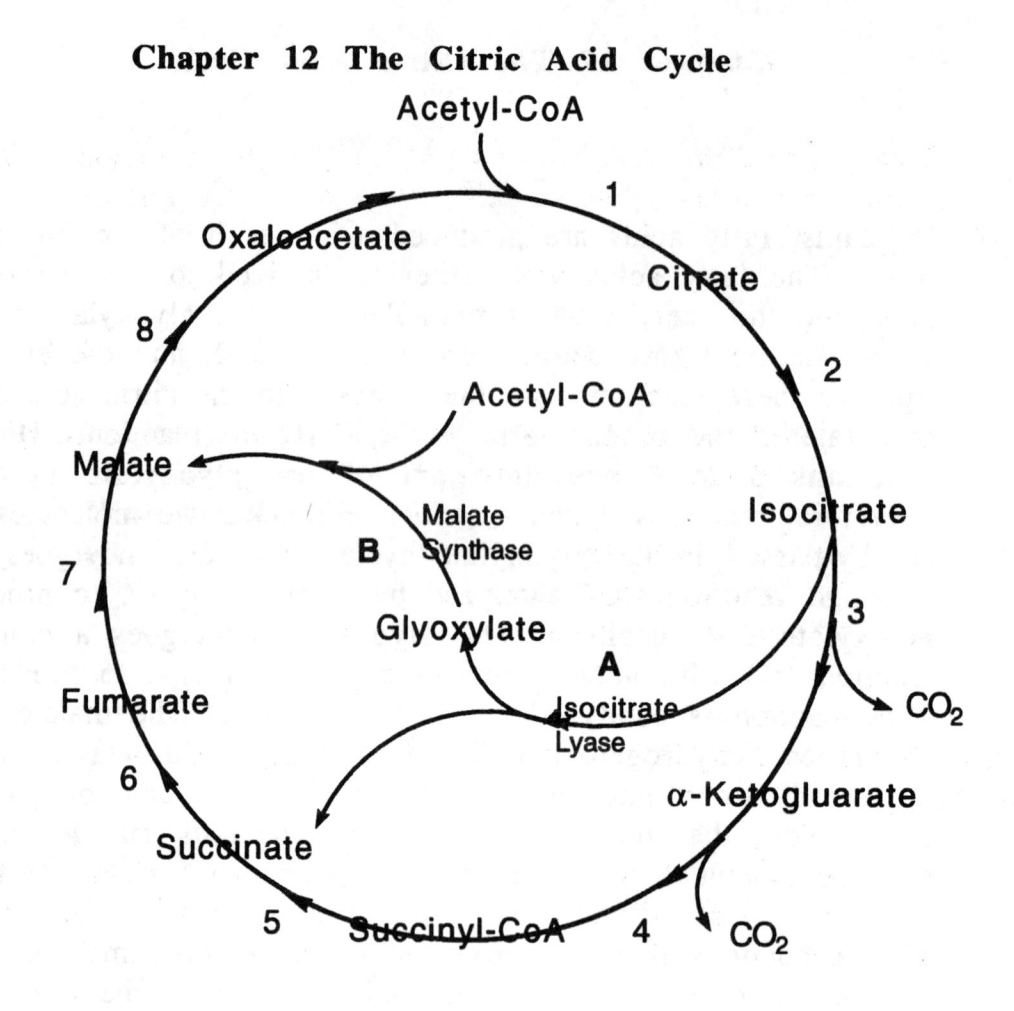

b. The glyoxylate cycle takes place in the glyoxysomes of plant cells.

c. Net carbohydrate synthesis from fat can be carried out in the glyoxylate cycle because there is a net synthesis of a four-unit carbon, succinate, in this cycle, This does not occur in the citric acid cycle. The succinate then shuttles to the mitochondria and into the citric acid cycle, where there is now an additional succinate per turn of the cycle. This succinate can then be used for carbohydrate synthesis.

d. The net equation for the glyoxylate cycle is:

2 acetyl-CoA + NAD$^+$ ----> succinate + NADH + H$^+$ + 2 CoA-SH

e. The conversion of pyruvate to acetyl-CoA, which is catalyzed by pyruvate dehydrogenase, is an irreversible reaction. The reverse reaction, conversion of acetyl-CoA to pyruvate, does not take place.

Chapter 12 The Citric Acid Cycle

f. Glyoxylate is the only molecule unique to the glyoxylate cycle. It is a two carbon molecule with the following structure:

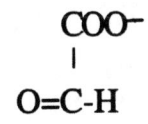

$$\begin{array}{c} COO^- \\ | \\ O=C-H \end{array}$$

9. a. Regulation of the citric acid cycle is accomplished by the following three enzymes:

   Citrate synthase
   Isocitrate dehydrogenase
   α-Ketoglutarate dehydrogenase

Notice that the regulation of the cycle occurs at reactions which have a large negative value for the standard free energy change (i.e., spontaneous reactions).

   b. In cells that have high levels of ATP and NADH, the citric acid cycle is inhibited since ATP and NADH act as allosteric inhibitors of the regulatory enzymes.

10. Cells in a highly active metabolic state require ATP to energetically drive biosynthetic reactions. Because of its continuous use, the level of ATP is low relative to the level of ADP. The presence of high levels of ADP stimulates glycolysis and the citric acid cycle in order to drive the synthesis of additional ATP. NADH is used at high rates since it directly generates ATP; therefore, its level is low relative to NAD^+.

The reverse is true for cells that undergo low metabolic activity or are in a metabolic resting state. In these cases, the ratio of ATP/ADP and NADH/NAD^+ are high.

13

Electron Transport and Oxidative Phosphorylation

The catabolic pathways for proteins, carbohydrates and lipids culminate in cellular respiration with **electron transport** and **oxidative phosphorylation** in the mitochondria. In the complete oxidation of one mole of glucose by molecular O_2, CO_2 and H_2O are produced with the liberation of 2823 kJ/mol of energy. In the process, 24 electrons are transferred indirectly from glucose to oxygen and much of the energy is captured and used for the synthesis of ATP from ADP and phosphate. The mechanism involves the exergonic and stepwise transfer of electrons from the reduced coenzymes, NADH and $FADH_2$, through the electron transport chain. Three major multienzyme complexes, which are located in the mitochondrial inner membrane, are key molecular assemblies in the respiratory chain. The use of respiratory inhibitors, which obstruct the transfer of electrons at specific sites in the chain, was instrumental in defining the order of the electron carriers in the membrane. The final enzyme in the chain, cytochrome oxidase, ultimately transfers the electrons to the final electron acceptor, molecular oxygen. The **chemiosmotic coupling theory** provides the most widely accepted

Chapter 13 Electron Transport and Oxidative Phosphorylation

mechanism to explain the coupling of this electron transport to the oxidative phosphorylation process. It proposes that the enzymes in the electron transport chain are also involved in pumping hydrogen ions from the matrix into the intermembrane space to create a **proton gradient**. It is this gradient which then provides the thermodynamic driving force for the phosphorylation of ADP by the multisubunit enzyme complex, ATP synthase.

Some Key Figures Revisited

Figure 13.2 Schematic representation of the electron transport chain, showing sites of proton pumping coupled to oxidative phosphorylation. FMN is the flavin coenzyme flavin mononucleotide, which differs from FAD in not having adenine nucleotide. CoQ is coenzume Q. Cyt b, cyt c_1, cyt c, and cyt aa_3 are heme-containing proteins cytochrome b, cytochrome c_1, cytochrome c, cytochrome aa_3, respectively.

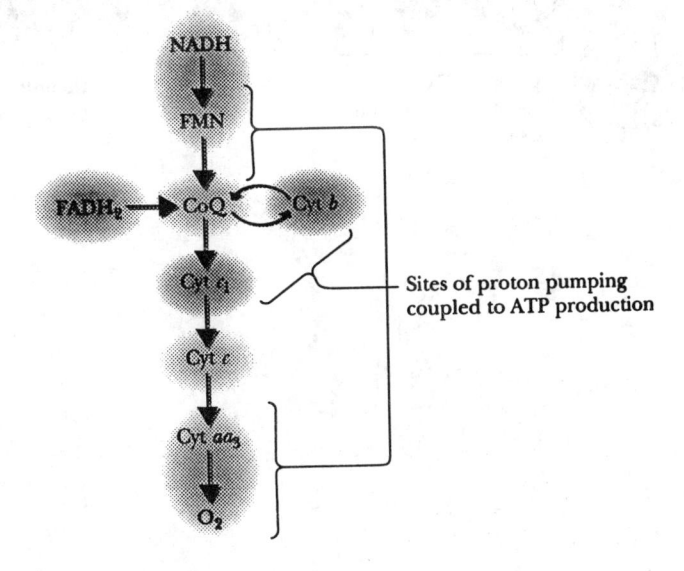

Chapter 13 Electron Transport and Oxidative Phosphorylation

Figure 13.4 The electron transport chain, showing the respiratory complexes. In the reduced cytochromes the iron is in the Fe(II) oxidation state, while in the oxidized cytochromes the oxygen is in the Fe(III) oxidation state.

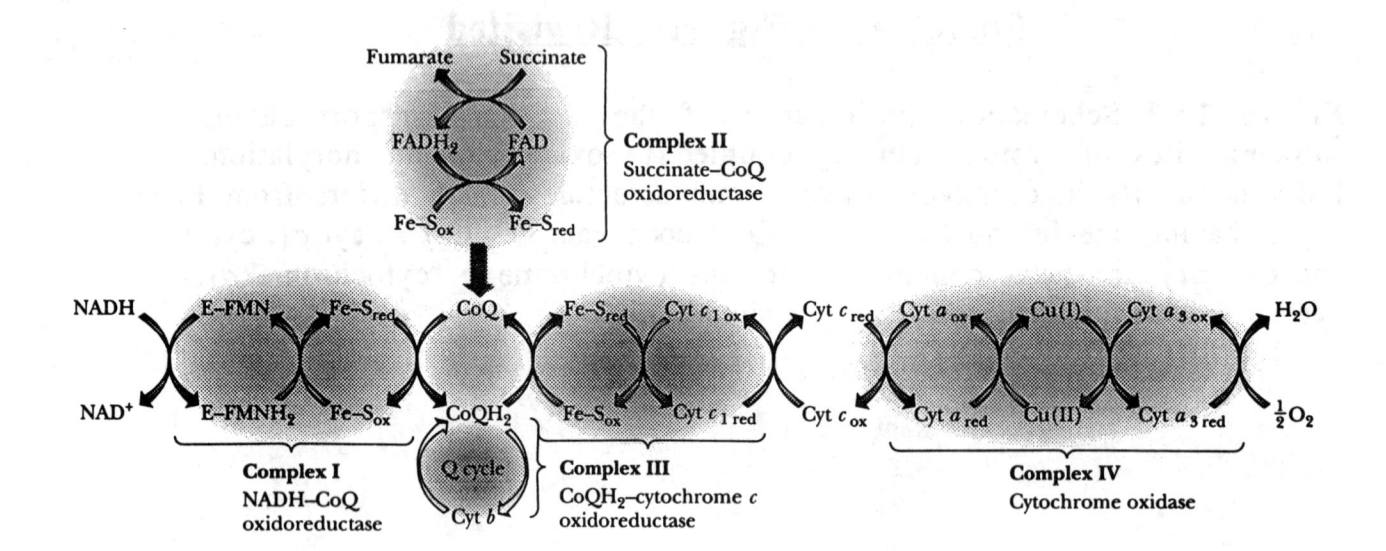

Chapter 13 Electron Transport and Oxidative Phosphorylation

Figure 13.6 The compositions and locations of respiratory complexes in the inner mitochondrial membrane, showing the flow of electrons from NADH to O$_2$. Complex II is not involved and not shown. NADH has accepted electrons from substrates such as pyruvate, isocitrate, α-ketoglutarate, and malate. Note that the binding site for NADH is on the matrix side of the membrane. Coenzyme Q is soluble in the lipid bilayer. Complex III contains two b-type cytochromes, which are involved in the Q cycle. Cytochrome c is loosely bound to the membrane, facing the intermembrane space. In Complex IV the binding site for oxygen lies on the side toward the matrix.

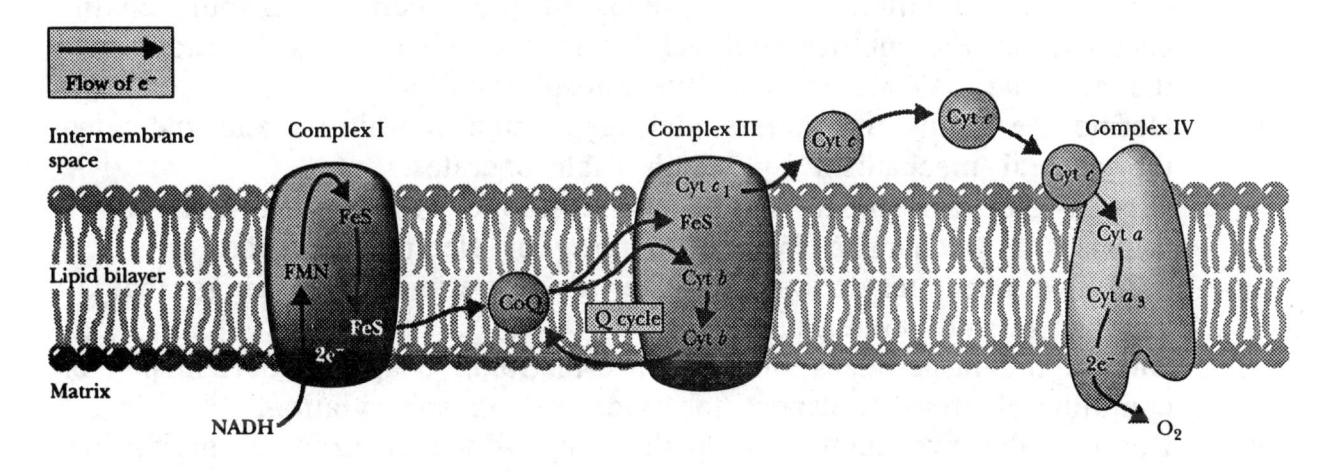

Figure 13.10 The creation of a proton gradient in chemiosmotic coupling. The overall effect of the electron transport reaction series is to move protons (H$^+$) out of the matrix into the intermembrane space, creating a difference in pH across the membrane.

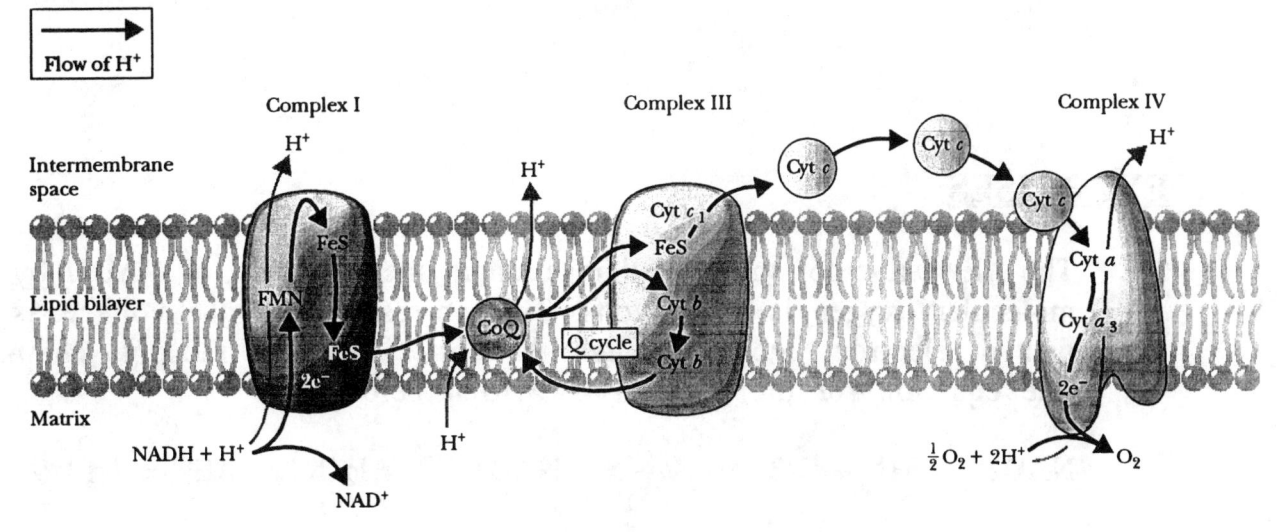

181

Chapter 13 Electron Transport and Oxidative Phosphorylation

CHAPTER OBJECTIVES

1. Indicate the cellular location where glycolysis, the citric acid cycle, electron transport and oxidative phosphorylation take place.
2. Specify the role of $FADH_2$, NADH and O_2 in the electron transport chain.
3. Name the three multienzyme complexes, the proteins, and the cofactors in each complex and the order in which the multienzyme complexes are found in the electron transport chain.
4. Indicate the electron carriers that serve to transfer electrons between the multienzyme complexes in the electron transport chain.
5. Characterize the multisubunit ATP synthase, identify its location in the cell and its role in oxidative phosphorylation.
6. Define the terms uncouplers and respiration inhibitors, and indicate the general mechanism by which each operates.
7. Discuss the basic tenets of the chemiosmotic hypothesis, which proposes a mechanism for the coupling of electron transport to oxidative phosphorylation. Indicate the experimental findings that support this hypothesis.
8. Outline the basic ideas in the conformational coupling mechanism for coupling electron transport to oxidative phosphorylation.
9. Describe the two shuttle mechanisms by which electrons from NADH are transported from the cytoplasm into the mitochondria.
10. Review the cytoplasmic and mitochondrial reactions involved in the complete oxidation of glucose and indicate the reactions which result in the production of ATP, NADH and $FADH_2$.

EXERCISES

1. The $\Delta E^{o'}$ value for the overall reaction in the oxidation of NADH by molecular oxygen is +1.14 volts. This corresponds to the voltage as a pair of electrons moves through the entire electron transport chain. The equation for the reaction is written below.

    $$NADH + H^+ + 1/2\ O_2 \longrightarrow NAD^+ + H_2O \qquad \Delta E^{o'} = 1.14\ V$$

Chapter 13 Electron Transport and Oxidative Phosphorylation

    a.   Calculate the $\Delta G^{o'}$ value and the equilibrium constant for this reaction. Use the relationships, $\Delta G^{o'} = -nF\Delta E^{o'} = -2.303 \log K_{eq}$, where n is the number of electrons transferred and F = 96,500 J/mol.

    b.   The oxidation of one mole of NADH in respiration produces 3 moles of ATP. What is the efficiency of the oxidative phosphorylation process under standard conditions?

2. Indicate which molecular species in the inner membrane of the mitochondria serves as proton pumps or channels.

3. Consult Figure 13.5 in the text for an abbreviated description of the energetics involved in the electron transport.
   a. Write the first reaction in the electron transport pathway, which involves NADH and CoQ.
   b. Indicate which reactants are the oxidized and reduced species, respectively.
   c. Indicate the oxidizing and the reducing agent in the reaction.
   d. Using the $\Delta E^{o'}$ value of +0.42 V, calculate the standard free energy value change for this reaction. This should verify the $\Delta G^{o'}$ value given in the Figure.

4. The molecule, 2,4-dinitrophenol, is an effective uncoupling agent.
   a. What effect does it produce in the electron transport chain and on the production of ATP?
   b. Suggest a possible mechanism for this process.
   c. If a laboratory rat is given 2,4-dinitrophenol, both its metabolism and body temperature increase. Explain this finding briefly.

5. The average male adult (160 lbs.) requires the consumption of food with the energy equivalent of about 2800 kcal of energy per day.
   a. If this energy is derived from the hydrolysis of ATP under standard conditions, calculate the number of moles and the number of grams of ATP which would be synthesized each day.
   b. It is found that there is about 50 g of ATP in the body. Considering this and the answer from part (a), suggest an explanation consistent with both findings.

Chapter 13 Electron Transport and Oxidative Phosphorylation

6. Indicate the cellular location of the following enzymes, molecules or processes. If the location is within the mitochondrion, specify whether it is in the matrix, inner membrane, inner membrane space or the outer membrane.
   a. Pyruvate dehydrogenase complex
   b. ATP synthase
   c. Citric acid cycle
   d. Triose-phosphate isomerase
   e. NADH
   f. ATP
   g. Malate
   h. Pentose phosphate pathway
   i. Cytochrome oxidase
   j. Succinate dehydrogenase

7. The overall equation for the complete oxidation of glucose by molecular oxygen is:

   $C_6H_{12}O_6 + 6 O_2 \longrightarrow 6 CO_2 + 6 H_2O$

   a. Write the two half-reactions for this redox process. These reactions involve the oxidation of glucose to CO_2 and the reduction of O_2 to H_2O. Show the number of electrons involved in each half-reaction.
   b. In the oxidation, indicate the electron acceptor molecule(s) that directly receive the electrons.
   c. Indicate the molecules or enzymes to which the electrons are transferred in the next phase of the process.
   d. Which molecule or enzyme transfers the electrons directly to molecular oxygen?

8. The $\Delta G^{o'}$ for the complete oxidation of glucose is - 2823 kJ of energy per mole of glucose.
   a. Calculate the percent of this energy that is efficiently converted to ATP in:
      i. Glycolysis
      ii. Pyruvate conversion to acetyl-CoA
      iii. Citric acid cycle
      iv. Electron transport and oxidative phosphorylation
   b. Calculate the energy conversion efficiency in the complete oxidation of glucose in a living cell.

Chapter 13 Electron Transport and Oxidative Phosphorylation

    c.   With the high negative value for $\Delta G^{o'}$ for this reaction, what can be predicted about the rate of the reaction?

9.   List the functions carried out by ATP synthase.

10.   Indicate if the following statements are true or false. If the statement is false, correct it.
    a.   The standard potential for a redox reaction is directly related to the standard free energy for the reaction.
    b.   The two species in the electron transport chain which are considered mobile electron carriers are coenzyme Q and cytochrome b.
    c.   Water is the terminal electron acceptor in the electron transport chain.
    d.   Electron transport is coupled to the production of ATP at three sites. These are the sites at which protons are pumped from the mitochondial matrix into the intermembrane space. The complexes involved in the pumping are referred to as Complexes II, III and IV.
    e.   Cytochrome a/a$_{3oxid}$ [Fe(III)] is a better oxidizing than cytochrome b$_{oxid}$.
    f.   All four of the protein complexes that make up the mitochondrial respiratory electron transport chain contain [Fe-S] proteins.
    g.   The reaction for the transfer of electrons from NADH to coenzyme Q has a $\Delta E^{o'}$ value of +0.42 volts. The standard reduction potential ($E^{o'}$) for the half-reaction for NAD$^+$/NADH is -0.32 volts. Therefore, the standard half-reaction potential for the reduction of coenzyme Q is +0.10 volts.

ANSWERS TO EXERCISES

1.   a.

$$\Delta G^{o'} = -nF\Delta E^{o'} = -(2)(96,500 \text{ J/mol})[1.14 \text{ V}] = \underline{-220 \text{ kJ/mol}}$$
$$\Delta G^{o'} = -2.303 \log K_{eq}$$

$$\underline{K_{eq} = 3.8 \times 10^{38}}$$

185

Chapter 13 Electron Transport and Oxidative Phosphorylation

    b.   The oxidation of one mole of NADH in the respiratory chain results in the production or synthesis of three moles of ATP. Since the $\Delta G^{o'}$ for the hydrolysis of ATP is -30.5 kJ/mol, the synthesis of three moles of ATP requires the input of 91.5 kJ/mol of energy. The energy efficiency for the process is:

$$\text{Efficiency} = [91.5 \text{ kJ/mol}]/[220 \text{ kJ/mol}] \times 100$$
$$\underline{\text{Efficiency} = 41.5 \%}$$

2. The three respiratory multienzyme complexes that serve as proton pumps are:
   1.   NADH-CoQ oxidoreductase (Complex I)
   2. $CoQH_2$-cytochrome c oxidoreductase (Complex III)
   3.   Cytochrome oxidase (Complex IV)
(The F_o functional subunit of ATP synthase acts as a proton channel.)

3.   a. $NADH + H^+ + CoQ = NAD^+ + CoQH_2$
     b.   The reactant which is the oxidized species is CoQ and the reduced species is NADH.
     c.   The oxidzing agent (species which will oxidize another species and as a result will be reduced itself) is CoQ. The reducing agent (species which will reduce another species and as a result be oxidized) is NADH.
     d.

$$\Delta G^{o'} = -nF\Delta E^{o'}$$
$$\Delta G^{o'} = -(2)(96.5 \text{ kJ/mol})(+0.42V)$$
$$\Delta G^{o'} = -81 \text{ kJ/mol}$$

4.   a.   The addition of 2,4-dinitrophenol (DNP) uncouples electron transport from the phosphorylation of ADP to produce ATP. That is to say, if DNP is added to a mitochondrial suspension that has an oxidizable substrate (such as succinate), O_2 will be consumed, yet no ATP will be produced. If the same experiment is carried out without the DNP, both O_2 consumption and ATP synthesis is observed.

     b.   DNP can exist in the protonated or unprotonated form.

Chapter 13 Electron Transport and Oxidative Phosphorylation

$$\text{DNP-OH} \rightleftharpoons \text{DNP-O}^- + H^+$$

At intracellular pH, DNP is in the dissociated form because of the pK_a of the phenolic hydrogen. If the DNP is in the vicinity of the inner membrane, where the pH is lower ($[H^+]$ is greater), DNP becomes protonated and reduces the free H^+ concentration in the inner membrane space. The protonation converts DNP from a hydrophilic, anionic species to a neutral, hydrophobic species. The latter species is soluble in the inner membrane and passes through it. In this way, it passively takes the protons from the inner membrane space into the matrix, bypassing the F_0 channel in the ATP synthase. As a result, ATP synthesis is uncoupled from the electron transport process. In the matrix where the pH is higher, the DNP dissociates the proton.

    c.   Since ATP levels will be decreased due to the uncoupling reaction, the regulation mechanisms in the cell will stimulate catabolic activity in order to promote the synthesis of ATP. However, this will not yield to ATP production because of the presence of the uncoupler. The body temperature will rise since the energy produced will be dissipated in the form of heat (and not used in ATP synthesis).

        Apparently, for a short period in the 1940s, the administration of sublethal doses of DNP was actually prescribed to produce weight lose in overweight individuals.

5.   a.   If the 2800 kcals of energy is used to synthesize ATP under standard condition (7.3 kcal/mol), the number of moles of ATP produced would be:

        [2800 kcal]/[7.3 kcal/ATP] = <u>384 moles ATP</u>
        The number of grams of grams of ATP would be:

        [384 moles][505 g/mole] = <u>193,920 g ATP</u>
    This corresponds to over 427 pounds of ATP!

Chapter 13 Electron Transport and Oxidative Phosphorylation

    b.   The ATP produced is hydrolyzed very quickly to ADP and phosphate and is then resynthesized thousands of times. The calculation shows that over 3878 times as much ATP is synthesized than is found in the body.

          [193,920 g ATP hydrolyzed]/ [50 g ATP in all cells] = 3878

6.
- **a.** Matrix of the mitochondrion
- **b.** Inner membrane of the mitochondrion
- **c.** Matrix of the mitochondrion, with succinate dehydrogenase being within the inner membrane
- **d.** Cytoplasm
- **e.** Throughout the cell
- **f.** Throughout the cell
- **g.** Cytoplasm and mitochondrion
- **h.** Cytoplasm
- **i.** Inner membrane of the mitochondrion
- **j.** Inner membrane of the mitochondrion

7.   **a.**   The half-reactions can be written independently as:

Oxidation: $C_6H_{12}O_6 + 6\ H_2O \longrightarrow 6\ CO_2 + 24\ H^+ + 24\ e^-$
Reduction: $6\ O_2 + 24\ e^- + 24\ H^+ \longrightarrow 12\ H_2O$

                $C_6H_{12}O_6 + 6\ O_2 \longrightarrow 6\ CO_2 + 6\ H_2O$

- **b.** The direct electron acceptors are NAD^+ and FAD, which are reduced to produce NADH and $FADH_2$, respectively.
- **c.** The electrons from the NADH are transferred to the first respiratory complex, the NADH-CoQ oxidoreductase, in the electron transport chain. $FADH_2$ is produced in the citric acid cycle by succinate dehydrogenase, which resides in the inner membrane of the mitochondria. The electrons do not formally leave this complex and are transferred directly to coenzyme Q in the electron transport chain.
- **d.** The cytochrome oxidase transfers the electrons to O_2.

Chapter 13 Electron Transport and Oxidative Phosphorylation

8. a.
   i. **Glycolysis**:
   $$C_6H_{12}O_6 + 2\,ADP + 2\,P_i + 2\,NAD^+ \longrightarrow$$
   $$2\,CH_3COCO_2^- + 2\,ATP + 2\,NADH + 2\,H_2O + 2\,H^+$$

   Glycolysis produces 2 ATPs.
   % of energy = [2(30.5 kJ/mol)]/[2823 kJ/mol] x 100 =
   <u>2.1% of total energy</u>

   ii. **Pyruvate Conversion to Acetyl-CoA**:
   $$2\,CH_3COCO_2^- + 2\,NAD^+ + 2\,CoAS\text{-}H \longrightarrow$$
   $$2\,Acetyl\text{-}CoA + 2\,NADH + 2\,CO_2$$

   Pyruvate conversion to acetyl-CoA does not produce any ATPs directly. There is <u>no energy contribution directly (i.e., ATPs) from this reaction.</u>

   iii. **Citric Acid Cycle**:
   $$2\,Acetyl\text{-}CoA + 4\,H_2O + 6\,NAD^+ + 2\,FAD + 2\,ADP + 2\,P_i \longrightarrow$$
   $$+ 4\,CO_2 + 6\,NADH + 6\,H^+ + 2\,FADH_2 + 2\,CoA\text{-}SH + 2\,GTP$$

   The citric acid cycle produces 2 GTPs directly (equivalent to 2 ATPs).
   % of energy = <u>2.1 % of total energy</u>. This is the same amount contributed by glycolysis.

   iv. **Electron Transport and Oxidative Phosphorylation**:
   $$10\,NADH + 10\,H^+ + 6\,O_2 + 2\,FADH_2 \longrightarrow$$
   $$10\,NAD^+ + 2\,FAD + 12\,H_2O$$

   The generation of ATP is derived from the transfer of the electrons from the energy-rich coenzymes into the electron transport chain and then to molecular oxygen. Each NADH oxidized generates 3 ATPs, while the oxidation of each $FADH_2$ results in the synthesis of 2 ATPs. Therefore, there is 34 ATPs produced.

   % energy = (34)(30.5 kJ/mol)/[2823 kJ/mol] x 100 =

   <u>36.7 % of total energy</u>

Chapter 13 Electron Transport and Oxidative Phosphorylation

b. The total energy produced in the form of ATP is:

Glycolysis:	61 kJ/mol
Citric acid cycle:	61 kJ/mol
Electron transport and oxidative phosphorylation	1037 kJ/mol
	1159 kJ/mol

% energy efficiency = [1159 kJ/mol]/ [2823 kJ/mol] x 100 = <u>41 %</u>

c. The free energy change is only an indication of the thermodynamic driving force for the reaction. There is no relationship between the $\Delta G^{o'}$ for a reaction and the rate of the reaction. Therefore, $\Delta G^{o'}$ values cannot be used to predict the rate of the reaction.

9. ATP synthase is the most complex assembly in the inner mitochondrial membrane. It contains many proteins and has two functional subunits, called F_o and F_1. It carries out three functions. It catalyzes: (1) the synthesis of ATP from ADP and phosphate; and (2) the reverse reaction, the hydrolysis of ATP to yield ADP and phosphate. Recall that an enzyme catalyzes both the forward and reverse reactions. These reactions are carried out by the F_1 subunit. The complex also (3) provides a channel for the translocation of protons from the inner membrane space to the matrix. This is accomplished by the F_o subunit.

10.
    a. True.
    b. False. The two species are coenzyme Q and cytochrome c.
    c. False. O_2 is the terminal electron acceptor in the electron transport chain.
    d. False. The complexes involved in pumping protons across the innner membrane are Complexes I, III and IV.
    e. True.
    f. False. Cytochrome oxidase does not contain [Fe-S] proteins.
    g. True.

14

Further Aspects of Carbohydrate Metabolism

Glucose is a major metabolic fuel. We have seen that a great deal of the energy from the oxidation of glucose generates ATP via glycolysis, the citric acid cycle and oxidative phosphorylation. Although organisms require a ready supply of glucose, they cannot store significant amounts of glucose monomers for future needs, such as in times of fasting or stressful situations. Organisms can, however, store glucose which is not immediately needed, in the form of the branched, polysaccharide, **glycogen**. Glycogen metabolism envelopes the biosynthetic pathways by which glycogen is synthesized from glucose and also the catabolic pathway by which stored glycogen in the liver and muscles is degraded to glucose. In addition, glucose also can be derived from non-carbohydrate precursors in the **gluconeogenesis pathway**. Although glycolysis and gluconeogenesis share many common enzymes and both pathways reside in the cytoplasm, they are distinct pathways, with different regulatory enzymes. In addition to the glycolysis and

Chapter 14 Further Aspects of Carbohydrate Metabolism

gluconeogenesis pathways, the activity of the regulatory enzymes in the anabolic and catabolic pathways of glycogen, must be controlled in a reciprocal manner. This control may be effected by allosteric interactions and/or covalent modifications. The **Cori cycle** demonstrates an example of metabolic interdependence and cooperation between organs. The **pentose phosphate pathway** provides yet another pathway in glucose metabolism. It can be viewed as an initial oxidative phase and a subsequent non-oxidative phase. Its primary function is to provide NADPH for biosynthetic reactions and to provide pentose phosphates. If the pentose phosphates are not needed, they are converted into intermediates, which enter the glycolytic pathway.

Some Key Figures Revisited

Figure 14.10 The Cori cycle. Lactate produced in muscles by glycolysis is transported by the blood to the liver. Gluconeogenesis in the liver converts the lactate back to glucose, which can be carried back to the muscles by the blood. Glucose can be stored as glycogen until it is degraded by glycogenolysis.

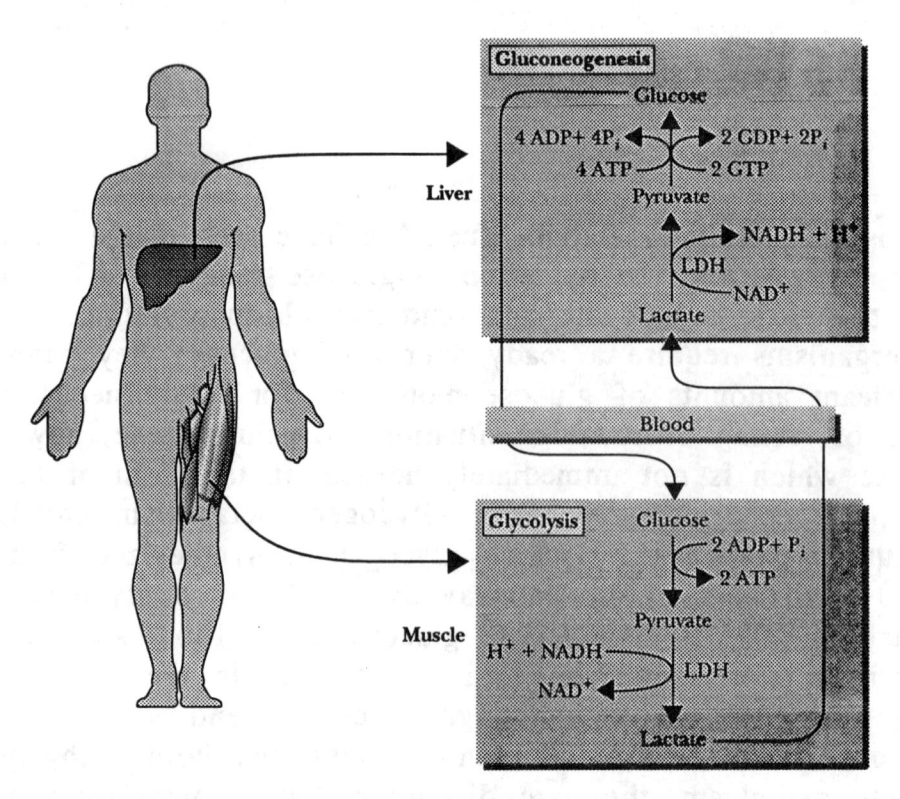

Chapter 14 Further Aspects of Carbohydrate Metabolism

Figure 14.14 Relationships between the pentose phosphate pathway and glycolysis. If the organism needs NADPH more than ribose-5-phosphate, the entire pentose phosphate pathway is operative. If the organism needs ribose-5-phosphate more than NADPH, the nonoxidative reactions of the pentose phosphate pathway, operating in reverse, produce ribose-5-phosphate.

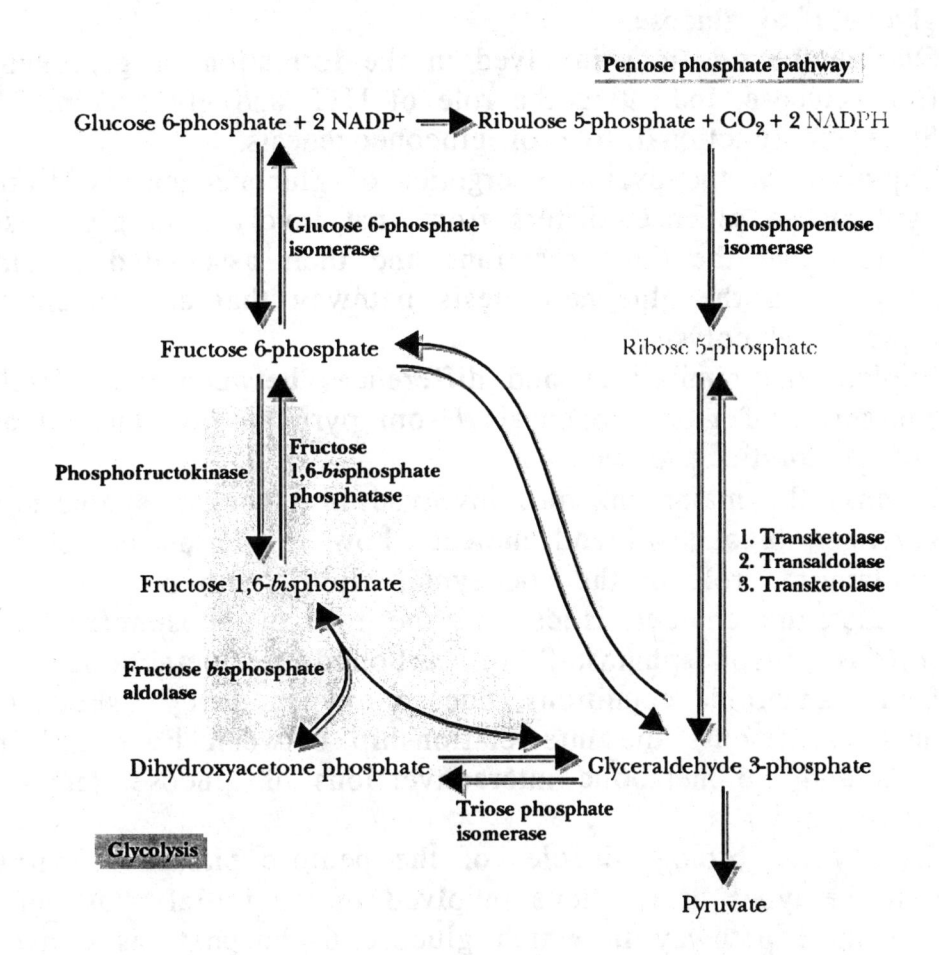

Chapter 14 Further Aspects of Carbohydrate Metabolism

Chapter Objectives

1. Characterize the structure and function of the polysaccharide, glycogen, and contrast these characteristics with that of starch and amylopectin.
2. Outline the individual steps involved in the conversion of glycogen to glucose.
3. Outline the reactions involved in the formation of glycogen from glucose, indicating the role of UTP and glycogenin.
4. State the functional role of gluconeogenesis.
5. Explain how the overall energetics of gluconeogenesis, from pyruvate to glucose, differs from that involved in glycolysis.
6. Write down the three reactions and their associated regulatory enzymes in the gluconeogenesis pathway that are different than those in glycolysis.
7. Explain the similarities and differences between the overall energetics of gluconeogenesis (from pyruvate to glucose) and the glycolytic pathway.
8. Discuss the major enzymes involved in glycogen synthesis and glycogen breakdown and indicate how regulation is achieved.
9. Explain the role of the coenzyme, biotin, with pyruvate carboxylase.
10. Discuss the characteristics and the role of phosphofructokinase-2 and fructose bisphosphatase-2 in carbohydrate metabolism.
11. Under anaerobic conditions, due to oxygen debt in skeletal muscles, discuss the interrelationship between liver and muscle cells and the metabolic interconversions of glucose, lactate and glycogen.
12. Specify the biological roles of the pentose phosphate pathway.
13. List the types of reactions involved in the initial phase of the pentose phosphate pathway in which glucose 6-phosphate is converted to ribulose 5-phosphate.
14. Indicate the general group transfer roles of transketolase and transaldolase in the nonoxidative phase of the pentose phosphate pathway.
15. Draw the structure for thiamine pyrophosphate and indicate the role of this coenzyme with transketolase.

Chapter 14 Further Aspects of Carbohydrate Metabolism

EXERCISES

1. Glycolysis is an exergonic pathway. The exact opposite pathway is not spontaneous. Explain how the conversion of pyruvate to glucose is a thermodynamic spontaneous process.

2. Muscle has an enormous need for ATP for motor action, especially under stressful conditions. This need for ATP leads to glycogen breakdown in the muscle. As a supplement, glycogen also can be broken down in the liver and the resulting glucose transported into the blood stream for delivery into the muscles. What would occur if there was a deficiency in glucose-6-phosphatase in the liver?

3. Write the reaction catalyzed by the following enzymes:
   a. glycogen phosphorylase
   b. glycogen synthase
   c. phosphorylase kinase
   d. pyruvate carboxylase

4. Give two examples of substrate cycling, naming the substrates and the enzymes involved.

5. Indicate the redox reactions that are in the "oxidative phase" of the pentose phosphate pathway. Write the reactions.

6. The primary functions of the pentose phosphate pathway are to generate: (1) NADPH for subsequent reductive biosynthesis; and (2) to generate ribose 5-phosphate for the synthesis of nucleotides.
   a. At what points in the pathway are these primary functions accomplished?
   b. Sum the individual reactions involved in the production of these two molecules and write the final net equation.

7. a. List the five coenzymes involved in reactions discussed in this chapter?
   b. For each coenzyme, name at least one enzyme that requires it for activity.
   c. What is the essential function of each coenzyme?

Chapter 14 Further Aspects of Carbohydrate Metabolism

8. What molecules are common to both glycolysis and the pentose phosphate pathway and thereby fix the interrelationship between the two pathways?

9. Cellular glutathione can reduce disulfide bonds in proteins. Write the general equation for this non-enzymatic reaction.

10. Write the reaction for the reduction of oxidized glutathione by NADPH.

ANSWERS TO EXERCISES

1. Glycolysis and gluconeogenesis differ in three steps in the pathways and, therefore, are not exactly reverse pathways of each other. Importantly, the three steps that differ provide the regulation for each pathway and are catalyzed by different enzymes. It is at these steps that the specific pathway gains its individual thermodynamic driving force. The other common steps (all these reactions are at, or near equilibrium; ΔG^o near zero) provide little or no thermodynamic driving force.

2. A deficiency of glucose-6-phosphatase would greatly limit the amount of available glucose since the conversion of glucose-6-phosphate to dlucose would be decreased. Under conditions in which glycogen breakdown in the liver is needed to provide a rapid supply of glucose, this breakdown could not be accomplished. The unphosphorylated, neutral glucose is needed to enter or leave the cell; phosphorylated glucose remains in the cell and cannot diffuse or be transported through the cell membrane. Therefore, the incomplete breakdown of glycogen to glucose in the liver would not be available to supplement the muscles in a stressful situation.

3.
   a. $[glycogen]_n$ residues + P_i ⇌ glucose-1-phosphate + $[glycogen]_{n-1}$ residues

   b. UDP-glucose + $[glycogen]_n$ ⇌ UDP + $[glycogen]_{n+1}$

   c. phosphorylase b + ATP ⇌ phosphorylase a + ADP

Chapter 14 Further Aspects of Carbohydrate Metabolism

    d.   pyruvate + CO_2 + ATP + H_2O ⇌ oxaloacetate + ADP + P_i

4. Substrate cycling refers to a reaction in which the opposing reactions are catalyzed by different enzymes. There are a number of examples of this, with three examples in the glycolysis/gluconeogenesis pathways in carbohydrate metabolism. Two prominent ones discussed in this chapter are;

a.   glucose + ATP $\xrightarrow{E_1}$ glucose-6-phosphate + ADP
$\xleftarrow{E_2}$

E_1 = glucokinase or hexokinase
E_2 = glucose-6-phosphatase

b.   fructose-6-phosphate (F6P) to fructose 1,6-bisphosphate (F1,6 BP)

F6P + ATP $\xrightarrow{E_1}$ F1,6 BP + ADP
$\xleftarrow{E_2}$

E_1 = phosphofructokinase
E_2 = fructose 1,6-bisphosphate

5. The first three reactions in the pentose phosphate pathway are considered the "oxidative phase". Of the three reactions, only the first (an oxidation) and the third (an oxidative decarboxylation) are oxidative, with $NADP^+$ serving as the oxidizing agent.

glucose 6-phosphate + $NADP^+$ ⇌ NADPH + H^+
                                           + 6-phosphoglucono-δ-lactone

6-phosphogluconate + $NADP^+$ ⇌ NADPH + H^+ + CO_2
                                           + ribulose 5-phosphate

6.   a.   The primary function is accomplished after the isomerization of ribulose 5-phosphate. The synthesis of NADPH has been accomplished in reactions 1 and 3 in the "oxidative phase."

Chapter 14 Further Aspects of Carbohydrate Metabolism

    b. The net reaction for the production of these molecules is:

glucose 6-phosphate + 2 NADP$^+$ ⇌ ribose 5-phosphate
$$+ CO_2 + 2\ NADPH + 2\ H^+$$

7. a, b, c.

Coenzyme	Enzyme(s)	Function
NAD$^+$	Many dehydrogenases (ex., glyceraldehyde 3-phosphate dehydrogenase)	Acts as an oxidizing agent (e$^-$ acceptor)
NADH	Lactate dehydrogenase	Acts as a reducing agent (e$^-$ donor)
NADP$^+$	Glucose 6-phosphate dehydrogenase	Acts as an oxidizing agent (e$^-$ acceptor)
TPP	Transketolase	Cleavage or formation of a carbon-carbon bond adjacent to a carbonyl carbon
Biotin	Pyruvate decarboxylase	A CO$_2$ carrier; Carboxylation or Transcarboxylation

8. glucose 6-phosphate, fructose 6-phosphate, glyceraldehyde 3-phosphate.

9.

```
 2 γ-glu-cys-gly  +  Protein   ⇌   γ-glu-cys-gly   +   Protein
         |               | |              |               |   |
         SH              S-S              S              SH  SH
                                          |
                                          S
                                          |
                                    γ-glu-cys-gly
```

Chapter 14 Further Aspects of Carbohydrate Metabolism

10.
```
γ-glu-cys-gly
     |
     S
     |     + NADPH + H⁺  ------->    2 glu-cys-gly  +  NADP⁺
     S                  <-------           |
     |                                    SH
γ-glu-cys-gly
```

Oxidized form   Reduced form

This reaction proceeds far to the right in most cells.

15

Lipid Metabolism

This chapter examines the metabolic processes associated with the catabolism and anabolism of a variety of lipid molecules. Triacylglycerols, being such highly reduced molecules, have the highest energy content of all the major nutrients. Fatty acid oxidation begins with the hydrolysis of the acylglycerol and the transport of the activated acyl group into the mitochondrion where the process of **β-oxidation** occurs. This cyclic pathway generates acetyl-CoA, which can be shuttled into the **Krebs cycle**, in addition to $FADH_2$ and NADH, which enter the electron transport chain and produce ATP by oxidative phosphorylation. Fatty acids with an odd number of carbon atoms or which have some degree of unsaturation are also oxidized by β-oxidation, but this requires additional reactions. In situations such as starvation or diabetes, an excess of acetyl-CoA is produced. The acetyl-CoA cannot enter the Krebs cycle and as a result, metabolites referred to as **ketone bodies** are formed. The biosynthesis of fatty acids is carried out in the cytosol by the successive addition of two-carbon units by the multienzyme complex, **fatty acid synthetase**. The anabolism of triacylglycerol, phosphoglycerol and sphingolipids is also briefly presented. The complex biosynthetic pathway for **cholesterol** is outlined, with a general overview presented for the conversion of cholesterol to bile

Chapter 15 Lipid Metabolism

salts and sex hormones. The role of serum cholesterol, low-density lipoprotein (LDL) particles and LDL receptors in cholesterol metabolism and in atherosclerosis is discussed.

CHAPTER OBJECTIVES

1. Indicate the cellular locations for the initial hydrolysis of acylglycerol and phosphatidyl choline, β-oxidation of fatty acids and fatty acid biosynthesis.
2. Outline the steps involved in the β-oxidation of fatty acids, starting with the hydrolysis of acylglycerol and including the transport mechanism that acyl groups utilize to enter the mitochondria.
3. Calculate the number of ATP molecules produced in the complete oxidation of a saturated or unsaturated fatty acid.
4. Consider the structures for the ketone bodies, the metabolic conditions in which they are produced and the reactions which describe their formation.
5. Describe the role of acetyl-CoA, malonyl-CoA and the acetyl-CoA carboxylase complex, fatty acid synthetase and acyl carrier protein (ACP) in fatty acid biosynthesis.
6. Summarize the important differences between the anabolic and catabolic pathways for fatty acids.
7. Outline the general aspects of the anabolism of acylglycerols and sphingolipids.
8. Name and characterize the major intermediates having C_2, C_6, C_5, C_{10}, C_{15}, and C_{30} units in the cholesterol biosynthetic pathway.
9. Characterize the structures for bile salts and the male and female sex hormones which are derived from cholesterol.
10. Describe the interrelationship between serum cholesterol levels, LDL and high-density lipoprotein (HDL) particles, LDL receptors and atherosclerosis.

Chapter 15 Lipid Metabolism
SPOTLIGHT

Atherosclerosis:
Cholesterol and The LDL Connection

Cholesterol is essential in higher organisms for both growth and viability. It serves as a component in all plasma membranes and is a precursor for the synthesis of bile salts, steroid hormones and vitamin D. It is primarily synthesized in the liver and to a lesser extent in the intestine, with about 800 mg/day produced in individuals on cholesterol-free diets. The major source of cholesterol in other cells is through the uptake of exogenous cholesterol in the form of serum LDL particles. The LDL particle (approximately 22 nm in diameter) provides the vehicle to transport insoluble cholesterol in the blood. When an animal cell requires cholesterol for membrane synthesis, it activates the synthesis of LDL receptor proteins and incorporates them in the cell membrane. The LDL particles actually contain esters of cholesterol, enveloped by a phospholipid layer, which contains cholesterol and at least one molecule of apoprotein B-100. Circulating LDLs bind to the cell-surface transmembrane proteins, the LDL receptors, by interacting specifically with the apoprotein B-100. The receptors cluster in regions of the membrane called coated pits, and after complexing with LDLs, the regions invaginate and undergo <u>receptor mediated endocytosis</u>, as outlined in Figure 15.30 in the text. The particles are metabolized in the cell to produce cholesterol, fatty acids and amino acids. The cholesterol is then used primarily in membrane synthesis.

Failure of the LDL particles to be taken up by the cells results in abnormally high levels of serum cholesterol. Because of the inherent insolubility of cholesterol and especially the cholesterol esters, which reside inside the LDL particle, these forms of cholesterol deposit on the inside of blood vessels after prolonged periods of high cholesterol levels. The cholesterol produces a plaque, which can ultimately block blood flow, resulting in a myocardial infarction, commonly known as a heart attack.

The failure of the LDLs to be taken into the cell can occur by a natural check-and-balance regulatory mechanism or,

Chapter 15 Lipid Metabolism

many times, as a result of mutations in the LDL receptors. In the former case, sufficient cholesterol within the cell suppresses endogenous cholesterol synthesis by inhibiting HMG-CoA reductase, the enzyme which catalyzes the committed step in cholesterol synthesis. Sufficient endogenous cholesterol also down regulates or suppresses the synthesis of LDL receptors, thereby restricting the uptake of additional cholesterol, beyond that needed by the cell. In this way, if there is a high level of cholesterol in the plasma, it remains there.

In addition, mutations in the LDL receptors can occur. These can result in any or all of the following ways:
1. Insufficient synthesis of receptors,
2. Receptors which are synthesized, but do not reach the plasma membrane,
3. Receptors that reach the membrane, but are defective and do not bind LDL particles,
4. Receptors that reach the membrane, bind LDL particles, but fail to cluster in the coated pits; therefore, they do not undergo endocytosis.

The medical concern in all these situations, of course, is the same: High cholesterol levels in the blood produce a greater risk of heart disease.

There are at least two ways to help lower this risk. One is to increase the level of high density lipoproteins (HDLs) in the plasma. It appears that a diet rich in polyunsaturated fatty acids, such as that found in fish and vegetable oils, can produce an increase in the HDL level. HDLs, by some unknown mechanism, decrease the level of LDLs. One known function of the HDLs is to remove cholesterol from the peripheral tissues and to return it to the liver. In the liver, the primary fate of cholesterol is to make bile salts, which are then secreted into the intestines and subsequently excreted from the body. The increased level of HDLs therefore helps to eliminate cholesterol from the body.

An alternate way to reduce serum cholesterol levels is to administer the drug, mevinolin (also called lovastatin), which is a potent competitive inhibitor of HMG-CoA reductase. By inhibiting cholesterol synthesis in the cells in this manner, the cell produces higher levels of LDL receptors and permits more LDL particles to enter the cells.

Chapter 15 Lipid Metabolism

This drug, therefore, reduces the levels of serum LDLs and the total cholesterol in the plasma.

EXERCISES

1. Acetyl-CoA is a molecule of central importance in cellular metabolism.
   a. Where is the site of acetyl-CoA synthesis in the cell?
   b. Indicate the precursors of acetyl-CoA in the cell.

2. Match the characteristics in the right column with the lipid molecules in the left column.

   a. Sphingomyelin   1. N-acylsphingosine

   b. Ganglioside   2. Complete hydrolysis produces a fatty acid, choline, phosphoric acid and sphingosine

   c. Sphingosine   3. Unsaturated 18 carbon chain containing an amine and two alcohol functional groups

   d. Ceramide   4. Formed by attachment of an oligosaccharide, containing sialic acid, to the primary alcohol group of ceramide

   e. Cerebroside   5. Glucose or other sugar is attached to the primary alcohol of ceramide

3. It is stated in Section 15.6 of the text that all 16 carbon atoms in palmitic acid are derived from the acetyl groups of acetyl-CoA. This statement can be extended to the other saturated fatty acids of relevance to biological systems. However, in Figure 15.11 in the text, it shows that the carboxylate group in malonyl-CoA, the key intermediate in fatty acid biosynthesis, is derived from the bicarbonate anion. Explain how all the carbon atoms in fatty acids can therefore be derived from only the acetyl groups in acetyl-CoA.

Chapter 15 Lipid Metabolism

4. The biosynthesis of triacylglycerol begins with the synthesis of glycerol 3-phosphate (G3P) from either glycerol or dihydroxyacetone phosphate. What are the subsequent steps and the enzymes in the synthesis of triacylglycerol?

5. In the metabolic pathway for cholesterol synthesis, the immediate precursor to mevalonate is β-hydroxy-β-methylglutaryl-CoA (HMG-CoA).
   a. Write the equation for the formation of HMG-CoA. Include the name of the enzyme that catalyzes this reaction.
   b. The first committed step in cholesterol synthesis is the synthesis of mevalonate. Write the equation for this reaction, including the important enzyme that catalyzes the reaction.
   c. Describe cholesterol in terms of the functional groups it exhibits, how many and what types of rings it has, and indicate the number of asymmetric carbons or optically active carbon centers in the molecule.

6. Consider a general acylglycerol in the cell. Outline the sequence of steps and the location of these events associated with the initial steps in fatty acid oxidation, up to the step at which the fatty acid enters the β-oxidation pathway.

7. In the β-oxidation of the acyl group, acetyl-CoA groups are formed in a stepwise manner. What is the chemical rational for the reactions that are involved in this process?

8. a. Name the three metabolic intermediates that are considered to be ketone bodies. Indicate which intermediates are actually ketones.
   b. Indicate under what conditions ketone bodies are produced in the cells.

9. Compare the number of ATP molecules formed by the oxidation of one mole of palmitic acid under (a) normal conditions and (b) severe starvation conditions. Consider, in this oversimplification, that in the extreme case, only an insignificant amount (none) of the acetyl-CoA enters the Krebs

Chapter 15 Lipid Metabolism

cycle. Therefore, only reduced coenzymes produced in the β-oxidation pathway enter the electron transport chain.

10. Indicate which of the terms (1-7) listed below are associated with the metabolic pathways;

    Fatty acid oxidation _____

    Fatty acid synthesis _____

    1. Acyl-CoA
    2. Malonyl-CoA
    3. Occurs in cytoplasm
    4. Uses NADPH
    5. Involves β-ketoacyl-ACP reductase
    6. In *E. coli,* pathway uses a multienzyme complex
    7. Uses FAD

11. Indicate whether the statements below are true or false. If the statement is false, correct it.
    a. In the fatty acid oxidation pathway, the enzyme, enol-CoA hydratase, requires a trans double bond arrangement between carbons 2 and 3.
    b. The β-oxidation pathway occurs in the cytosol.
    c. All of the carbons in cholesterol are derived from acetyl-CoA.
    d. The oxidation of saturated fatty acids requires more reactions in their metabolic break down than the oxidation of unsaturated fatty acids.
    e. Before fatty acid synthesis can occur, acetyl-CoA must be transported out of the mitochondria and into the cytosol. This is accomplished by a transport mechanism in which oxaloacetate is transported across the membrane.
    f. ACP, which is part of the multienzyme complex called fatty acid synthetase, binds to acetyl or acyl groups by a thioether linkage.
    g. CTP is required for the biosynthesis of phosphatidylethanolamine in both bacteria and mammals.

Chapter 15 Lipid Metabolism

ANSWERS TO EXERCISES

1. a. Acetyl-CoA is produced in the mitochondria of the cell.
   b. It can be formed from (1) decarboxylation of pyruvate, (2) β-oxidation of fatty acids, and (3) from the catabolism of certain amino acids (Chapter 17).

2. a. 2
   b. 4
   c. 3
   d. 1
   e. 5

3. Malonyl-CoA is produced from acetyl-CoA and HCO_3^- in a reaction catalyzed by acetyl-CoA carboxylase. Two of the three carbons in the malonyl group are derived from the acetyl group and one from the bicarbonate anion. The biosynthesis of fatty acids, however, involves the successive addition of only two carbon units, so that only the carbons from the acetyl group are added to the growing chain. The carbon derived from the bicarbonate anion is eliminated in this reaction as a molecule of CO_2.

4. The reactions are the following:

   $$G3P + Acyl\text{-}CoA \xrightarrow{GPAT} Lysophosphatidate + CoA\text{-}SH$$

   G3P = Glycerol 3-phosphate
   GPAT = Glycerol phosphate acyl transferase

   $$Lysophosphatidate + Acyl\text{-}CoA \xrightarrow{GPAT} Phosphatidate + CoA\text{-}SH$$

   $$Phosphatidate + H_2O \xrightarrow{Phosphatidate\ Phosphatase} 1,2\text{-Diacylglycerol} + P_i$$

Chapter 15 Lipid Metabolism

$$\text{1,2-Diacylglycerol} + \text{Acyl-CoA} \xrightarrow[\text{Acyl Transferase}]{\text{Diacylglycerol}} \text{Triacylglycerol} + \text{CoA-SH}$$

5. a.

Acetoacetyl-CoA   Acetyl-CoA   β–Hydroxy-β-methylglutaryl-CoA

$$\underset{\text{CH}_3\text{-C-CH}_2\text{-C-S-CoA}}{\overset{\text{O} \quad\quad \text{O}}{\| \quad\quad \|}} + \underset{\text{H}_3\text{C-C-S-CoA}}{\overset{\text{O}}{\|}} + \text{H}_2\text{O} \xrightarrow{E} {}^-\text{O}_2\text{C-CH}_2\text{-}\underset{\underset{\text{CH}_3}{|}}{\overset{\overset{\text{OH}}{|}}{\text{C}}}\text{-CH}_2\text{-}\overset{\overset{\text{O}}{\|}}{\text{C}}\text{-S-CoA}$$

$$+ \text{ CoA-SH} + \text{H}^+$$

E = Hydroxymethylglutaryl-CoA synthase

b.

β–Hydroxy-β-methylglutaryl-CoA   Mevalonate

$${}^-\text{O}_2\text{C-CH}_2\text{-}\underset{\underset{\text{CH}_3}{|}}{\overset{\overset{\text{OH}}{|}}{\text{C}}}\text{-CH}_2\text{-}\overset{\overset{\text{O}}{\|}}{\text{C}}\text{-S-CoA} + 2\text{ NADPH} \xrightarrow{E} {}^-\text{O}_2\text{C-CH}_2\text{-}\underset{\underset{\text{CH}_3}{|}}{\overset{\overset{\text{OH}}{|}}{\text{C}}}\text{-CH}_2\text{-CH}_2\text{OH}$$

$$+ \text{ 2 NADP}^+ + \text{CoA-SH}$$

E = Hydroxymethylglutaryl-CoA reductase (HMG-CoA reductase)

c. Cholesterol contains only two functional groups, an alcohol at C3 and an olefin linkage between carbons 5 and 6 in the 6-membered rings. There are 4 fused rings, one of which is a 5-membered, cyclopentane ring and three 6-membered rings, two cyclohexane rings and one cyclohexene ring.

There are eight asymmetric carbons in cholesterol located at positions 3, 5, 8, 9, 10, 13, 17 and 20. The numbering system for cholesterol is shown in the accompanying figure.

208

Chapter 15 Lipid Metabolism

Numbering System in Cholesterol

6. In the cytosol, a lipase cleaves the acylglycerol to produce glycerol and fatty acids. The fatty acid reacts with an ATP to form an acyl-adenylate intermediate and pyrophosphate, with the latter being hydrolyzed to phosphates. The acyl-adenylate then is activated by formation of a thioester bond with CoA-SH. This latter reaction is catalyzed by acyl-CoA synthetase. The acyl-CoA is transported through the outer mitochondrial membrane, but cannot pass through the inner membrane into the matrix where the remaining reactions associated with oxidation are carried out. At this point, the acyl group is enzymatically transferred to carnitine by carnitine acyltransferase. The acyl carnitine crosses the inner membrane and then the acyl group is transferred back to CoA-SH. The activated acyl group is now ready to enter into the β-oxidation pathway in which two-carbon units are successively cleaved off the carboxyl end.

7. Starting with a saturated acyl group with n carbons, the aim of the cycle is to produce acetyl units in a step-wise manner. These units will be in the form of acetyl-CoA, which can then undergo further oxidation in the Krebs cycle. The stepwise process will involve converting the acyl group into a molecule which can be attacked by CoA-SH to produce an acetyl-CoA and an (n-2) acyl group. The process will continue until all acetyl-CoA units are produced.

Chapter 15 Lipid Metabolism

To convert the hydrocarbon portion of the acyl group to a ketone, an organic chemist would reduce a (C-C) bond first to an olefin linkage, then hydrate it to form a secondary alcohol and finally oxidize this alcohol to a ketone (refer to summary of organic reactions in Chapter 2 of this problems book). This is the same rational and strategy taken in the cell.

8. a. The three metabolic intermediates that have historically been referred to as ketone bodies are,

   Acetoacetate
   Acetone
   β-Hydroxybutyrate

The first two are ketones, while the third molecule is a hydroxy acid.

b. Ketone bodies are formed during starvation conditions and in cases of diabetes. In the former case, there are no glucose or glycogen stores to generate energy via the Krebs cycle and oxidative phosphorylation. Large amounts of acetyl-CoA are produced from fatty acid oxidation, but the acetyl-CoA cannot enter the Krebs cycle because of a depletion of oxaloacetate. This cannot be generated because there is no pyruvate from glycolysis. The buildup of acetyl-CoA therefore results in the reaction of two molecules of acetyl-CoA to form acetoacetyl-CoA and ketone bodies.

Diabetes mellitus, in some ways, mimics starvation. Although there can be a plentiful supply of glucose in the blood, the cells are "metabolically starving" since the glucose cannot enter the cells to undergo conversion to pyruvate and then acetyl-CoA. The outcome in both starvation and diabetes is similar, with high levels of ketone bodies being produced in both cases.

9. Palmitic acid contains 16 carbon atoms. It will require 7 cycles of the β-oxidation pathway to yield 8 moles of acetyl-CoA. This will produce 7 moles of $FADH_2$ and 7 moles of NADH. The 8 moles of acetyl-CoA produced from 1 mole of palmitic acid will enter the Krebs cycle. For each mole of acetyl-CoA, one mole of $FADH_2$ and 3 moles of NADH are produced, in addition to one mole of GTP. The NADH and $FADH_2$ enter the electron transport chain and produce ATP by oxidative phosphorylation.

Chapter 15 Lipid Metabolism

Overall ATP produced:
1. <u>β-oxidation</u>:   7 FADH$_2$ (Enters electron transport chain)
                           7 NADH   (Enters electron transport chain)
                           8 acetyl-CoA   (Enters the Krebs cycle)

2. <u>Krebs cycle</u>:   8 FADH$_2$ (Enters electron transport chain)
                           24 NADH   (Enters electron transport chain)
                           8 GTP

3. <u>Electron transport and oxidative phosphorylation</u>:
     Total FADH$_2$ 15 FADH$_2$ x 2 ATP =   30 ATP
     Total NADH   31 NADH   x 3 ATP =   93 ATP
     (From Krebs cycle)   8 GTP =   8 ATP equivalents
                                                             ———
                                                             131 ATP

Deduct 2 ATP equivalents
due to the activation step.   - 2 ATP
                                                             ———
                               Total   <u>129 ATP molecules</u>

A net total of 129 molecules of ATP are produced for each molecule of palmitic acid which is completely oxidized

In the case of severe starvation and considering the simplifications noted above, only the 7 NADH and the 7 FADH$_2$ produced in the β-oxidation of palmitic acid would yield ATPs. The acetyl-CoA molecules would go to produce ketone bodies. Within this approximation, the number of molecules of ATP produced is:

                 7 NADH x 3 ATPs   = 21 ATP
                 7 FADH$_2$ x 2 ATPs   = <u>14 ATP</u>
                                                     35 ATP

                 Activation   - <u>2 ATP</u>
                    Total   <u>33 ATP molecules</u>

             [33 ATP/129 ATP] x 100 = <u>25.5 %</u>

Less than 26% of the normal energy would be produced under these conditions.

Chapter 15 Lipid Metabolism

10.
- a. 1, 6, 7
- b. 1, 2, 3, 4, 5

11.
- a. True.
- b. False. It occurs in the mitochondria
- c. True.
- d. False. This is correct for unsaturated fatty acid synthesis.
- e. False. Citrate is the molecule which is transported across the membrane.
- f. False. This involves a thioester linkage.
- g. True.

16

Photosynthesis

The sun is the ultimate source of all biological energy. **Photosynthesis**, which is the process that captures this radiant energy and converts it into chemical energy, can be regarded as two major processes. The first process involves a series of reactions in which solar energy is absorbed by chlorophyll molecules in the chloroplasts of green plants. This energy is then used to power endergonic reactions which produce carbohydrates and molecular O_2 from the simple precursors, H_2O and CO_2. Two structurally distinct **Photosystems**, I and II, located in the **thylakoid membrane** of the chloroplast, are the light collection centers in which the excited chlorophyll molecules trigger a series of electron transfer reactions. These initial reactions are referred to as the "light reactions" and are used to produce O_2, ATP and NADPH. The reactions are strikingly similar to those observed previously in the mitochondria, where electron transport takes place by using membrane bound proton-pumping oxidoreductases. In this process, a proton gradient is established across the thylakoid membrane and this electrochemical force is used to drive the production of ATP. The second major process in photosynthesis involves the so-called "dark reactions," which use ATP and NADPH to reduce CO_2 to carbohydrates in the stroma of the chloroplasts. The fixation of CO_2 takes place via the

Chapter 16 Photosynthesis

metabolic loop called the **Calvin cycle**. In this process, ribulose 1,5-bisphosphate is the initial reactant and is regenerated, while glucose is derived from glyceraldehyde 3-phosphate, which leaves the cycle.

Some Key Figures Revisited

Figure 16.6 Electron flow in Photosystems I and II. The energy needed to transfer electrons from H_2O to $NADP^+$ is provided by the absorption of light by Photosystems I and II (vertical up arrows). After each absorption of light, the electrons can then flow "downhill" (diagonal down arrows). Photophosphorylation of ADP to yield ATP is coupled to the electron transport chain that links Photosystem I to Photosystem II. (Chl is chlorophyll; Phe is pheophytin; PQ is plastoquinone; PC is plastocyanin). The electron carriers that mediate the transfer of electrons from H_2O to Photosystem II include a manganese-containing protein and a protein with an essential tyrosine residue referred to as component Z.

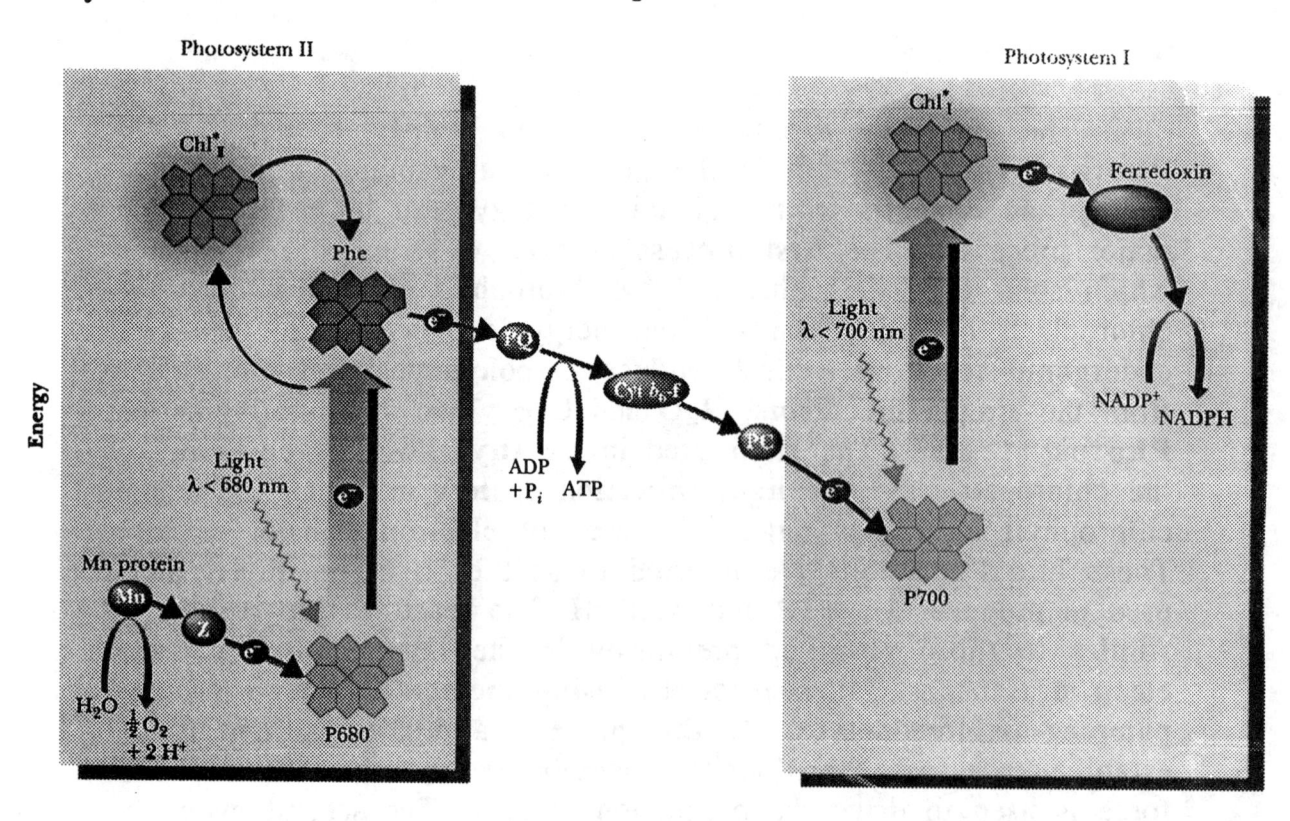

Chapter 16 Photosynthesis

Figure 16.19 The complete Calvin cycle, showing the regeneration of ribulose 1,5-bisphosphate. Note that when glyceraldehyde 3-phosphate is formed, it (or dihydroxyacetonephosphate, to which it is easily converted) can have all four possible fates. The possible pathways are numbered. (See Figure 16.18 in the text for the balanced equation.)

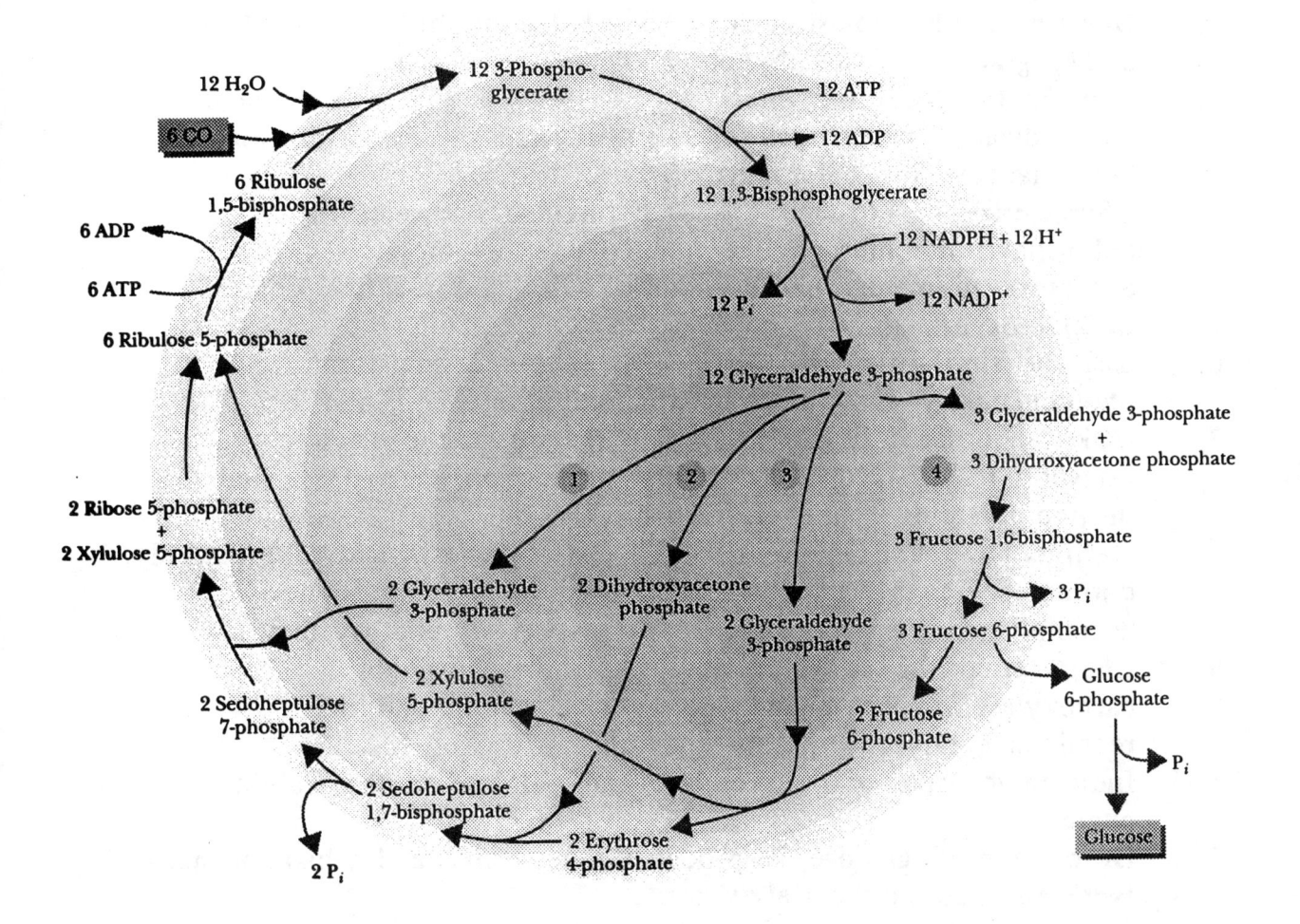

215

Chapter 16 Photosynthesis

CHAPTER OBJECTIVES

1. Write the simple equation that describes photosynthesis and the more descriptive equation that includes ATP and NADPH.
2. Write the separate reactions for the light and the dark photosynthetic reactions.
3. Describe a chloroplast, characterizing the stroma, thylakoid disks, grana and the thylakoid membrane structure.
4. Indicate the locations where photosystems I and II are found and indicate where the light reactions and the dark reactions take place in the chloroplasts.
5. Characterize the structures and distinguish the roles of chlorophyll *a*, chlorophyll *b*, and bacteriochlorophyll *a*, accessory pigments, photosynthetic units and reaction centers in photosynthesis.
6. List the similarities and differences in the composition, photochemistry and functions for photosystems I and II.
7. Make an ordered list of molecular species in the electron transport reactions involved in the production of NADPH and O_2 in photosystems I and II, respectively.
8. Outline how ATP is produced in photosynthesis and, alternatively, outline the manner in which ATP production can also be accomplished by cyclic electron transport in photosystem I.
9. Characterize the enzyme, ribulose 1,5-bisphosphate carboxylase:oxygenase and specify its location and the reactions catalyzed by this enzyme.
10. Indicate the role of transketolase and aldolase in the Calvin cycle.
11. Describe how glucose is produced and ribulose 1,5-bisphosphate is regenerated in the Calvin cycle.
12. Outline the C_4 pathway for CO_2 fixation and describe its advantage in tropical plants.

EXERCISES

1. Select the molecules or functions (1-15) which are associated with items (a-e) in the listing below.

Chapter 16 Photosynthesis

a. Photosystem I _____

b. Photosystem II _____

c. Calvin Cycle _____

d. Chlorophyll _____

e. Accessory pigments _____

f. Reactions connecting _____
   PS I and II

1. Chloroplasts
2. λ < 700 nm
3. Thylakoid membrane
4. ATP synthesis
5. O_2 synthesis
6. Transketolase
7. Phytol group
8. NADPH synthesis
9. Pheophytin
10. Stroma
11. Plastoquinone
12. Ferredoxin
13. Mn containing enzyme
14. 3-Phosphoglycerate
15. Xylulose 5-phosphate isomerase

2. Ribulose 1,5-bisphosphate carboxylase:oxygenase (RuBisCO) is the most abundant enzyme in nature and probably the most abundant protein. It has been estimated that there are about 40 million **tons** of it in the world.
   a. Where is RuBisCO found in the chloroplast?
   b. How large is it and how many subunits make up this enzyme?
   c. Indicate which reactions RuBisCO catalyzes.

3. Indicate whether the following statements are true or false. If the statement is false, correct it.
   a. Tropical plants circumvent the Calvin cycle by synthesizing glucose by the C_4 pathway.
   b. Chl_{II}^+ is a strong oxidant in photosystem II and can reduce water. Chl_{II}^+ is reduced by the transfer of electrons from the H_2O.
   c. In photosystem I, soluble ferredoxin is the acceptor molecule for the direct electron transfer from Chl_I^*.

Chapter 16 Photosynthesis

    d. Electrons pass from photosystem I to photosystem II by the electron transport chain.

    e. The primary role of an accessory pigment, such as pheophytin, is to absorb light.

    f. The energy of light at 680 nm is greater than that at 700 nm.

    g. Photosystem I is involved with the production of NADPH, while photosystem II carries out the splitting of water to produce molecular oxygen.

    h. Transketolases are enzymes which are involved in the transfer of C_2 units.

    i. Photorespiration is known to be a very efficient process in its use of ATP and NADPH.

    j. Cyclic electron transfer produces ATP under cellular conditions when the ratio of NADPH/NADP$^+$ is low.

4. Although the Calvin cycle and the Krebs cycle are very different, they exhibit some common characteristics. Make a comparison of the two cycles.

5. Both photosystems I and II (PS I and II) are involved in redox reactions. Indicate which species and reactions in these photosystems exhibit the following characteristics:

    a.   i. The strongest oxidizing agent in PS II: _____
         ii. The species oxidized by this agent: _____
         iii. Primary oxidation half-reaction carried out in PS II:

       _____

    b.   i. The strong reducing agent in PS I: _____
         ii. The chain of species reduced by this agent: _____
         iii. Primary reduction half-reaction carried out in PS I:

       _____

    c. The products of the above reactions which take part in the dark reactions: _____

    d. The overall equation for the light reaction:

       _____

Chapter 16 Photosynthesis

    e.   Explain how ATP is generated as a result of this reaction.

    f.   Indicate the fate of the products formed in the light reactions.

6. Consider the light reactions in photosynthesis. Complete the table or statements below:
   a. How many photons are absorbed by PS I and by PS II for each electron flowing from H_2O to $NADP^+$?

   <u>Number of Photons</u>

   PS I   _____
   PS II   _____
   Total   _____

   b. To form one molecule of O_2:
       i. _____ (#) molecules of H_2O are oxidized.
       ii. _____ (#) electrons must flow from two molecules of H_2O to reduce _____ (#) molecules of $NADP^+$ in order to produce _____ (#) molecules of NADPH.

   c. In the entire process of producing one molecule of O_2, a total of _____ (#) photons must be absorbed, with _____ (#) absorbed by each photosystem.

7. Summarize the processes occurring in PS I and PS II with respect to the overall light reaction.

8. Consider the Calvin cycle:
   a. How many complete turns of the cycle are required to produce one molecule of glucose?
   b. Does the regeneration of ribulose 1,5-bisphosphate require more or less molecules of ATP than the synthesis of glucose?
   c. After six complete turns of the Calvin cycle, how many molecules of glyceraldehyde 3-phosphate are produced and how many are used in the synthesis of glucose?

9. Complete the following statements concerning the C_4 pathway for CO_2 fixation.
   a. CO_2 enters the plant leaf through the _____ cells, which are the outer most cells.
   b. The CO_2 reacts with _____ to form _____.

Chapter 16 Photosynthesis

  c. Oxaloacetate is converted to malate by the enzyme _____.
  d. _____ is the molecule which leaves the _____ cells to enter the _____ cells, where the reactions in the Calvin cycle take place.
  e. The CO_2 which enters the Calvin cycle is derived from the decarboxylation of _____ , with _____ being the other product.

ANSWERS TO EXERCISES

1. a. 1, 3, 8, 12
   b. 1, 2, 3, 5, 13
   c. 1, 6, 10, 14, 15
   d. 1, 2, 3, 7
   e. 1, 3
   f. 1, 3, 4, 9, 11

2. a. RuBisCO is located on the stromal surface of the thylakoid membrane.
   b. RuBisCO has a molecular weight of 560 kDaltons, with eight large subunits, of molecular weight 55 kDaltons and eight smaller subunits, of molecular weight 15 kDaltons Interestingly, the gene for the larger protein is encoded by a chloroplast gene, while the gene for the smaller subunit, is encoded by the DNA within the nucleus. This is also a case in which one of the subunits (55 kDaltons) exhibits a catalytic function, while the smaller (15 kDaltons) subunit is thought to be involved in a regulatory role.
   c. RuBisCO is the abbreviation for <u>ribulose 1,5-bisphosphate carboxylase:oxygenase</u>; therefore, it serves as both a carboxylase and an oxygenase.

The carboxylase function is involved in the fixation of CO_2 which diffuses into the stroma. Ribulose 1,5-bisphosphate, a phosphorylated, five-carbon ketose (monosaccharide), reacts with

Chapter 16   Photosynthesis

CO₂ to form an unstable six-carbon intermediate, which splits apart to produce two molecules of 3-phosphoglycerate.

RuBisCO is a fascinating and genuinely unique enzyme. The single active site in RuBisCO, which carries out the fixation of CO₂, is also the same active site that facilitates the oxidation of ribulose 1,5-bisphosphate to 3-phosphoglycerate and phosphoglycolate. This reaction initiates the process known as photorespiration. Note that this enzyme is classified as an <u>oxygenase</u> because it carries out an oxidation in which O₂ is not only involved, but becomes incorporated into the product, 3-phosphoglycolate. This reaction diverts ribulose 1,5-bisphosphate from the Calvin cycle under conditions of high O₂ levels and low CO₂ levels.

3.
- a. False. The C₄ pathway leads to the Calvin cycle.
- b. False. Chl$_{II}^+$ is reduced by the transfer of electrons from water, while H₂O is oxidized to O₂ in the process.
- c. False. Chl$_I^*$ transfers its excited electron to chlorophyll *a*.
- d. False. The transfer of electrons is from PS II to PS I.
- e. False. The accessory pigments are involved in both light-trapping reactions and electron transfer processes.
- f. True. Recall that $E = h\upsilon = h(c/\lambda)$.
- g. True.
- h. True. It transfers the two-carbon glycoaldehyde unit.
- i. False. It is very inefficient in the utilization of ATP and NADPH.
- j. False. Cyclic electron transfer produces ATP when the ratio of NADPH/NADP⁺ is high.

4.

Characteristic	Calvin Cycle	Krebs Cycle
1. Type of cycle	Reductive	Oxidative
2. ATP and NAD(P)⁺	Uses ATP and NADPH	Generates ATP, NADH and FADH₂
3. General function	Carbohydrate synthesis	Central pathway in integration of metabolism; catabolism and anabolism of carbohydrates, proteins and lipids
4. Entering unit	CO₂ (C₁ unit)	Primarily acetyl (C₂ units) from acetyl-CoA
5. Regenerated molecule	Ribulose 1,5-bisphosphate	Oxaloacetate

Chapter 16 Photosynthesis

5. a.   i. Chl_{II}^+
      ii. H_2O
      iii. $2 H_2O \longrightarrow 4 H^+ + 4 e^- + O_2$

   b.   i. Chl_I^*
      ii.   chlorophyll a, bound ferredoxin, soluble ferredoxin, $NADP^+$
      iii. $NADP^+ + 2 e^- + H^+ \longrightarrow NADPH$

   c.   ATP and NADPH

   d. $2H_2O + 2 NADP^+ \longrightarrow O_2 + 2 NADPH + 2 H^+$

   e.   There are 2 hydrogen ions produced in the overall light reaction for every 2 molecules of water oxidized. In addition, a number of protons are produced in the reactions that connect PS II and PS I. Although the total number of protons is not firmly established, it is estimated that there are about two to three protons produced for each electron (i.e., 8-12 H^+ per molecule of O_2 formed). These protons are produced in the thylakoid space (See Figure 16.11 in the text.) and the net result is a pH gradient across the thylakoid membrane. The pH in the thylakoid space is lower than that in the stroma and the difference has been determined to be as great as 3.5 pH units. The protons can only pass through the membrane and into the stroma by going through the membrane-bound proton translocating ATP synthetase complex and thereby providing the thermodynamic driving force for the phosphorylation of ADP to produce ATP.

   f. O_2 diffuses out of the chloroplasts. NADPH and the generated ATP are utilized in the dark reactions for the synthesis of glucose.

6. The half-reaction for the oxidation of water shows that for every two molecules of water oxidized, one molecule of O_2, four electrons and four protons are produced.

$$2 H_2O \longrightarrow O_2 + 4 H^+ + 4 e^-$$

Chapter 16 Photosynthesis

a. <u>Number of Photons</u>

PS I	1
PS II	1
Total	2

b. i. 2
   ii. 4, 2, 2, 2
c. 8, 4

7. Upon absorption of light (λ < 700 nm) in PS I, an electron is excited and therefore expelled from the reaction center. This excited electron "flows down" a chain of electron carriers to NADP$^+$ and reduces it to NADPH. This process leaves an "electron hole" or a missing electron in PS I. This, in turn, is filled with an electron expelled from PS II by the capture of a photon of light with a wavelength < 680 nm. This electron is transferred from PS II to the electron transport chain and then to P700 in PS I. PS II now has an "electron hole" and this is filled by electrons arising from the oxidation of water. In the process, protons are produced to generate the proton gradient which drives the production of ATP. The molecular oxygen that is produced is released from the chloroplast.

8. a. Six turns of the Calvin cycle are required to produce one molecule of glucose.
   b. Regeneration of ribulose 1, 5-bisphosphate requires the hydrolysis of more molecules of ATP than does the synthesis of glucose.
   c. Twelve molecules of glyceraldehyde 3-phosphate are generated. Only two molecules of glyceraldehyde 3-phosphate are used for glucose synthesis.

9. a. mesophyll
   b. phosphoenolpyruvate, oxaloacetate
   c. malate dehydrogenase
   d. Malate, mesophyll, bundle sheath
   e. malate, pyruvate

The following figure briefly outlines the C$_4$ pathway.

223

Chapter 16 Photosynthesis

Mesophyll Cell	Bundle Sheath Cell
$CO_2 \rightarrow CO_2$ → Oxaloacetate -----> Malate ------>	Malate → CO_2 → Calvin Cycle
Phosphoenolpyruvate <--Pyruvate <---- (AMP ATP)	Pyruvate

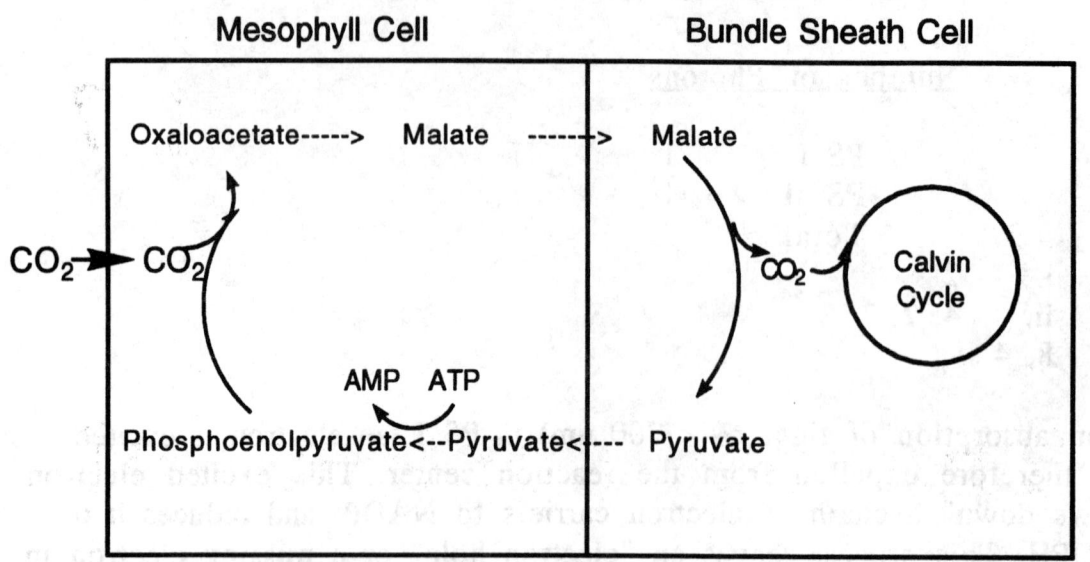

Abbreviated Features of the C_4 Pathway in Plants

17

The Metabolism of Nitrogen

Nitrogen metabolism encompasses the biosynthesis and degradation of amino acids, porphyrins, purine and pyrimidine nucleotides and nitrogen-containing hormones and coenzymes. The ultimate source of nitrogen in biological molecules is inorganic N_2, which must be "fixed" into a useable form such as NH_3. Important aspects of amino acid anabolism involve transamination and one-carbon transfer reactions, which require pyridoxal phosphate and tetrahydrofolate coenzymes, respectively. The catabolism of amino acids involves the elimination of nitrogen by way of the **urea cycle** and the metabolism of the residual carbon skeleton by the **citric acid cycle**. Purine nucleotides are synthesized *de novo* by the stepwise building of the ring systems of the purine base on ribose 5-phosphate or alternatively, by the more energy efficient salvage pathway. The anabolism of pyrimidine nucleotides starts with the synthesis of the pyrimidine ring before it is attached to the ribose 5-phosphate. The pathway by which deoxyribonucleotides are synthesized from the ribonucleotides can be inhibited by molecules that are used in cancer chemotherapy.

Chapter 17 The Metabolism of Nitrogen

Some Key Figures Revisited

Figure 17.5 The relationship between amino acid metabolism and the citric acid cycle.

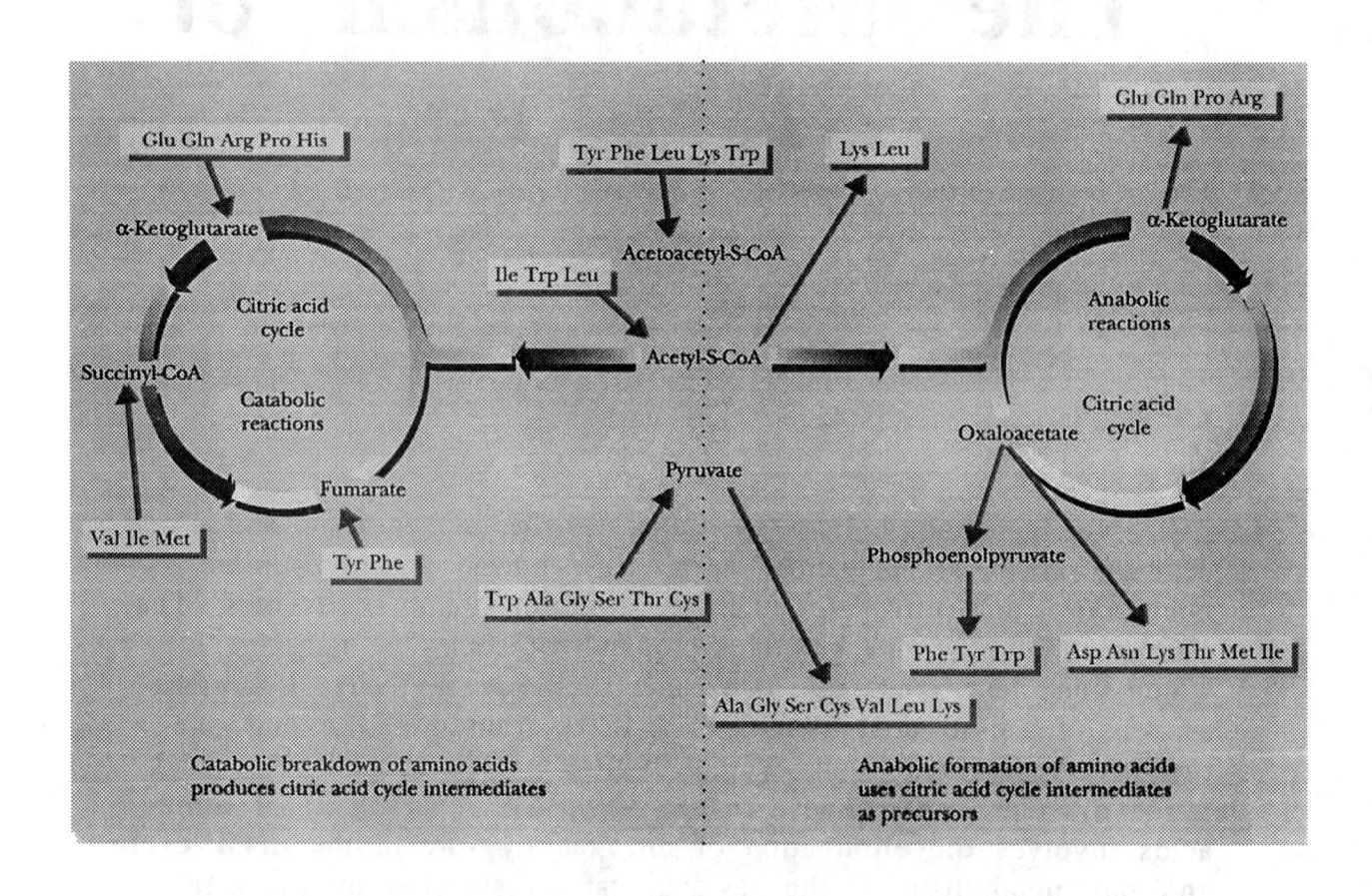

Chapter 17 The Metabolism of Nitrogen

Figure 17.15 The urea cycle and some of its links to the citric acid cycle.

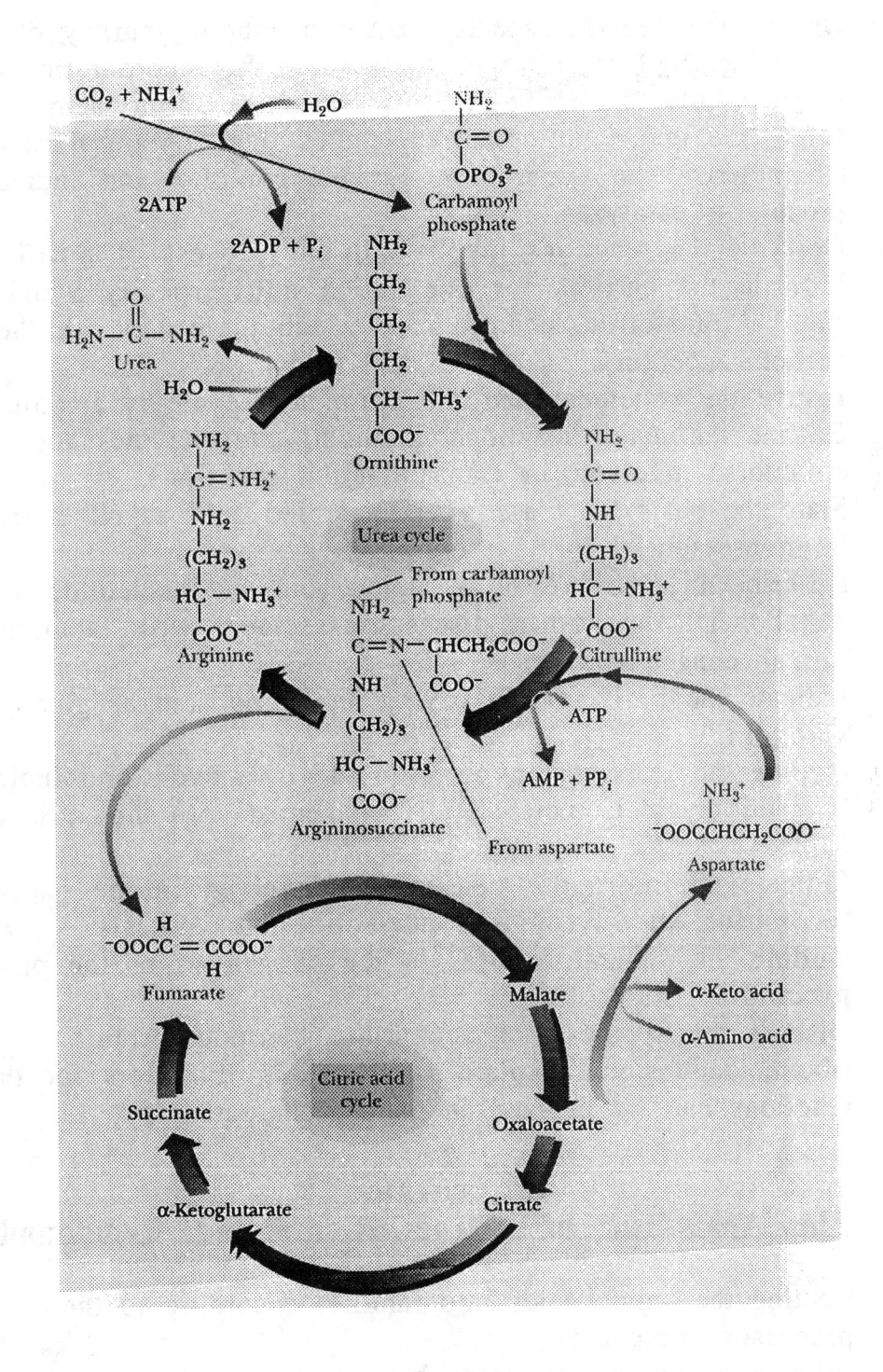

Chapter 17 The Metabolism of Nitrogen

CHAPTER OBJECTIVES

1. Outline the general aspects for the metabolic journey of inorganic N_2 from nitrogen fixation to the incorporation of nitrogen into simple biomolecules such as amino acids.
2. Name the major biological molecules that contain nitrogen.
3. Characterize the nitrogenase enzyme complex and indicate the process it catalyzes.
4. Describe the reactions involved in the process of nitrification.
5. Indicate the carriers for one-carbon units, specify which groups can be transferred, and indicate which molecules are the one-carbon acceptors.
6. Justify the statement that the citric acid cycle is amphibolic.
7. Outline the major reactions, including the enzymes and cofactors, involved in transamination reactions.
8. State the basis for classifying an amino acid as either essential or non-essential.
9. Describe an inborn error in amino acid and nucleotide metabolism, the human diseases associated with them and the linical consequences.
10. Indicate the basis for classifying an amino acid as glucogenic or ketogenic.
11. Outline the reactions involved in the urea cycle and indicate how this cycle is metabolically connected with the citric acid cycle.
12. Indicate the molecular conversions involved in the pathways for purine and pyrimidine nucleotide biosynthesis.
13. Outline the general features in the catabolism of the purine and pyrimidine nucleotides.
14. Discuss the synthesis of deoxyribonucleotides from ribonucleotides and explain why specific inhibitors for these reactions can be utilized in cancer chemotherapy.

Interchapter B
The Anabolism of Nitrogen-Containing Compounds

1. Catalog the amino acids into families according to their precursor molecules.
2. Indicate which amino acids have shikimate as an intermediate; indicate the central importance of chorismate in this pathway.

Chapter 17 The Metabolism of Nitrogen

3. Outline the major reactions associated with the biosynthesis of porphyrins and heme.
4. Outline the stepwise reactions in the biosynthesis of inosine-5'-monophosphate.

EXERCISES

1. The nitrogen fixation reaction for the conversion of inorganic N_2 to NH_3 has a standard free energy of -33 kJ/mol. The bond energy for the triple bond in N_2 is enormous, at 940 kJ/mol.

$$N_2 + 3 H_2 \rightleftharpoons 2 NH_3$$

The fixation of N_2 is not only vital to biological systems, but it is an important industrial process in the manufacture of ammonia for fertilizers. To make this reaction go to the right to a significant extent, the reaction is carried out **industrially** at very high temperature and pressure. Microorganisms, however, cannot survive in such extreme conditions. The reaction is accomplished in multiple steps by the enzyme, nitrogenase, and requires the input of about 12 molecules of ATP.

   a. Write 3 distinct reactions for the stepwise reduction of N_2, assuming that in each step two reducing equivalents (one equivalent being either a H or a H^+ and an e^-) are added.
   b. With a ΔG^o of -33 kJ/mol, which indicates a spontaneous reaction, why are such drastic industrial conditions needed and why are such elaborate enzyme complex necessary to produce NH_3 in nitrogen-fixing bacteria?

2. The following molecules are involved in the conversion of serine to glycine: pyridoxal phosphate, serine hydroxymethylase and tetrahydrofolate.
   a. Indicate the role of each of these molecules in the conversion process.
   b. What type of bond is formed between the amino acid and the pyridoxal phosphate?

Chapter 17 The Metabolism of Nitrogen

3. The citric acid cycle is said to be amphibolic.
   a. What does this expression mean?
   b. Indicate the four intermediates in this cycle that serve as entry points for the catabolism of amino acids.
   c. Indicate the two intermediates in the citric acid cycle which serve as intermediate precursors for the anabolism of amino acids.

4. Name the three important one-carbon carriers encountered in metabolic processes.

5. a. Indicate why an amino acid is designated as glucogenic, ketogenic or both glucogenic and ketogenic.
   b. Explain why organisms that have a glyoxylate cycle do not have ketogenic amino acids.

6. a. There are two "molecular-links" between the urea cycle and the citric acid cycle. Name these two molecules and indicate the nature of the linkage.
   b. It is stated that the urea and citric acid cycles are linked by the formation and breakdown of a single intermediate in the urea cycle. What is this metabolic intermediate? Explain briefly in light of your answer in (a).
   c. How many molecules of ATP are involved in the synthesis of urea and how many high energy bonds are hydrolyzed?
   d. Urea has two nitrogen atoms. What are the molecular sources of these nitrogens which are eliminated in the form of urea?

7. a. What are the substrates for the salvage reactions in nucleotide metabolism?
   b. Indicate the reactions and the enzymes used in the salvage synthesis of purine nucleotides.

8. Aspartate transcarbamoylase (ATCase) is one of the most extensively studied regulatory enzymes.
   a. Is this a simple or a multisubunit enzyme?
   b. Does it follow Michaelis-Menten kinetics?
   c. Write the equation for the reaction that it catalyzes.
   d. What does it mean for a reaction to be the "committed step" of a metabolic pathway?

Chapter 17 The Metabolism of Nitrogen

    e.   What are the allosteric effectors for aspartate transcarbamoylase?

9. Ribonucleotides are synthesized and then converted to deoxyribonucleotides.
   a. Write the reaction associated with this conversion. Indicate the enzyme and the proteins associated with this reaction.
   b. What type of reaction is this?
   c. What are the common abbreviations for the nucleoside diphosphates produced in this reaction?
   d. How are thymine nucleotides produced from uracil nucleotides? Write the equation for this reaction, including the enzyme and any cofactors.

10. Explain how an individual with lead poisoning may develop anemia.

11. Indicate whether the statements below are true or false. If a statement is false, correct it.
    a. Nitrate reductase and nitrite reductase are essential enzymes in the nitrification process.
    b. When excess amino acids are consumed by an organism, beyond that needed for the synthesis of proteins, the amino acids are stored and then used as needed.
    c. α-Ketoglutarate is the major acceptor for amino groups from the other amino acids.
    d. Pyridoxal phosphate, which serves as a coenzyme in transamination reactions, is bound to the enzyme by a Schiff base linkage.
    e. The carbon unit involved in the one-carbon transfers with tetrahydrofolate cofactors is bonded to either the N^5 or the N^{10} atoms or both nitrogen atoms.
    f. In purine nucleotide biosynthesis, the purine ring system is assembled on the ribose 5-phosphate, while in pyrimidine biosynthesis, the pyrimidine ring is synthesized before being attached to the ribose 5-phosphate.
    g. The salvage pathway for the biosynthesis of purine nucleotides is complex and requires a great deal of

Chapter 17 The Metabolism of Nitrogen

energy. As a result, it is used in the cell as only a last resort for purine nucleotide synthesis.

h. Lesch-Nyhan syndrome is associated with a deficiency of the enzyme, hypoxanthine-guanine phosphoribosyltransferase (HPRT).

i. Inhibitors of folate reductase, such as fluorouracil, are widely used in cancer chemotherapy.

j. Ribonucleotides are precursors for deoxyribonucleotides.

ANSWERS TO EXERCISES

1. a. The distinct steps in this reaction are:
   1. $N_2 + 2 H^+ + 2 e^- \longrightarrow H-N=N-H$
   2. $H-N=N-H + 2 H^+ + 2 e^- \longrightarrow H_2N-NH_2$
   3. $H_2N-NH_2 + 2 H^+ + 2 e^- \longrightarrow 2 NH_3$

   b. The ΔG^o for a reaction represents the thermodynamic driving force for the reaction and is an indication of whether the reaction goes to the right to generate products. The ΔG^o value, however, does not indicate the rate at which this reaction will take place. Although this reaction is spontaneous and will go to the right, there are effectively no products formed under standard conditions. The reaction has an extremely high energy of activation which must be overcome for ammonia to form. This energy barrier can be overcome by increasing the temperature and pressure, in line with Le Chatelier's principle. Alternatively, the reaction can be accomplished enzymatically, in which case the activation energy is significantly lowered.

2. a. Serine hydroxymethylase is the enzyme responsible for the catalytic conversion of serine to glycine. Pyridoxal phosphate is the coenzyme required for transamination reactions. (It also takes part in amino acid decarboxylations, racemizations and a number of modifications of the amino acid side group). The tetrahydrofolate is the coenzyme involved in the transfer of the single carbon unit.

Chapter 17 The Metabolism of Nitrogen

    b.   A Schiff base linkage is formed between the amino acid and the pyridoxal phosphate.

3.   a.   An amphibolic cycle is one involved in both catabolism and anabolism.

    b.   α-Ketoglutarate, succinyl-CoA, fumarate and oxaloacetate serve as entry points into the Krebs cycle for the metabolic breakdown of some amino acids.

    c.   α-Ketoglutarate and oxaloacetate serve as intermediate precursors for the biosynthesis of some amino acids.

4.   The three one-carbon carriers are, biotin (CO_2); tetrahydrofolate (methyl, methylene, methenyl, formimino or formyl groups); and S-adenosylmethionine (methyl groups).

5.   a.   After the removal of the nitrogen from the amino acids, the carbon skeletons of the normal amino acids can be degraded to one or more of the following metabolic intermediates:

1. acetoacetyl-CoA
2. acetyl-CoA
3. pyruvate
4. α-ketoglutarate
5. succinyl-CoA
6. fumarate
7. oxaloacetate

    The carbon skeletons of the **glucogenic** amino acids are degraded to the intermediates, 3 through 7. All of these can be used as precursors for glucose. The carbon skeletons for the **ketogenic** amino acids are converted to either acetoacetyl-CoA or acetyl-CoA, which then can be converted to ketone bodies. Since animals lack a metabolic pathway for the **NET** conversion of acetoacetate or acetyl-CoA to glucogenic precursors, no **NET** synthesis of glucose or carbohydrate is possible.

    b.   Plants and some microorganisms have a glyoxylate cycle. This cycle makes the **NET** synthesis of glucose synthesis possible for all the amino acids. In the glyoxylate cycle, there is a net synthesis of oxaloacetate from two molecules of acetyl-CoA. The oxaloacetate formed can be

Chapter 17 The Metabolism of Nitrogen

used in gluconeogenesis. However, in the Krebs cycle, the conversion of acetyl-CoA to oxaloacetate is accompanied by the elimination of two molecules of CO_2. As a result of this elimination, the **NET** synthesis of carbohydrate from acetyl-CoA (a C_2 group) is impossible.

The following figure shows the seven common metabolic intermediates and the amino acids (the carbon skeletons), which are degraded to each intermediate. Although there is general agreement regarding the degradation pathway for the amino acids, there is not universal agreement about the assignments for amino acids in the diagram below.

Chapter 17 The Metabolism of Nitrogen

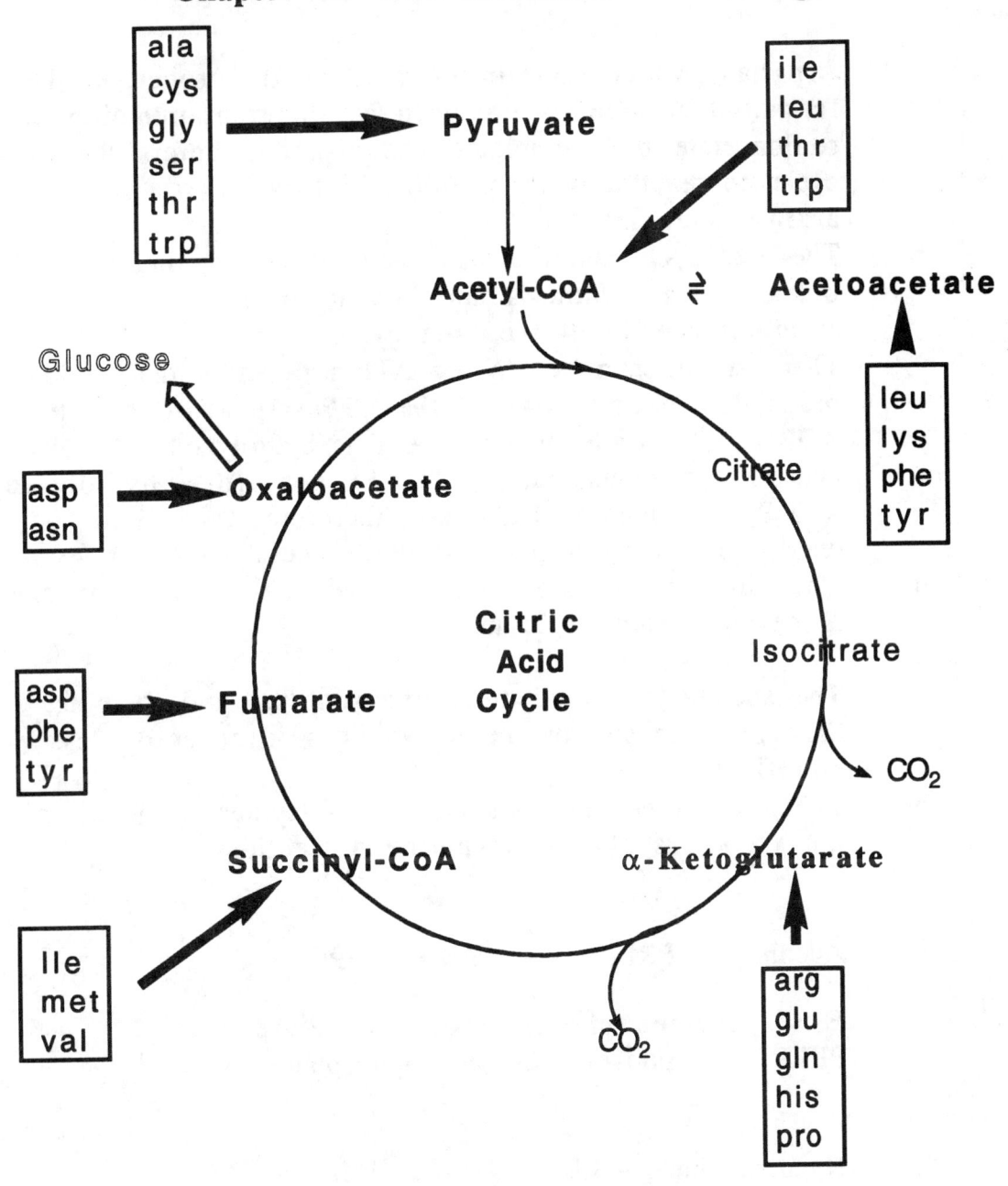

Amino acids are degraded to 7 metabolic intermediates

6. a. The two molecules that link the urea and citric acid cycle are fumarate and aspartate. Fumarate is a product, which exits the urea cycle, in the reaction that splits argininosuccinate. The other product of this reaction is arginine, the immediate precursor of urea (and ornithine). Fumarate can then shuttle into the citric acid cycle, where it is one of the intermediates.

235

Chapter 17 The Metabolism of Nitrogen

      Aspartate, which contributes the second nitrogen in the formation of urea, is produced by the transamination of oxaloacetate by glutamate. The aspartate enters the urea cycle by reaction with citrulline to form argininosuccinate.

- b. The urea cycle and the citric acid cycle are, therefore, linked by the synthesis and breakdown of argininosuccinate in the urea cycle.
- c. There are three molecules of ATP utilized in the production of urea. Two of the ATPs are hydrolyzed to ADP and P_i, while the third ATP is hydrolyzed to AMP and PP_i (pyrophosphate). The PP_i is further hydrolyzed to 2 P_i (phosphate). There are, therefore, four high energy bonds hydrolyzed in the process.
- d. One nitrogen comes from NH_4^+, while the second nitrogen is derived from aspartate.

7.
- a. The substrates are adenine, hypoxanthine and guanine. Salvage synthesis for the pyrimidines is generally less significant.
- b. There are three reactions and two enzymes involved in the salvage synthesis of purine nucleotides.

    i. Adenine + PRPP $\xrightarrow{E_1}$ AMP + PP_i

    E_1 = adenine phosphoribosyltransferase
    PPRP = phosphoribosylpyrophosphate

    ii. Hypoxanthine + PRPP $\xrightarrow{E_2}$ IMP + PP_i

    iii. Guanine + PRPP $\xrightarrow{E_2}$ GMP + PP_i

    E_2 = hypoxanthine-guanine phosphoribosyltransferase

8.
- a. ATCase is a multisubunit enzyme, as are all regulatory enzymes.
- b. Allosteric enzymes do not obey Michaelis-Menten kinetics.

Chapter 17 The Metabolism of Nitrogen

c.

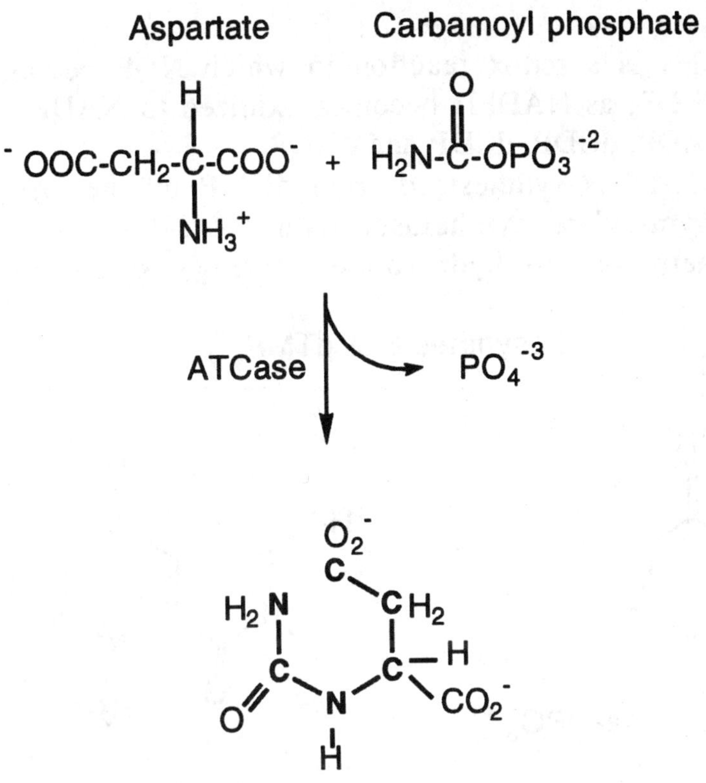

Aspartate + Carbamoyl phosphate

$$^-OOC-CH_2-\underset{NH_3^+}{\overset{H}{C}}-COO^- + H_2N-\overset{O}{\overset{\|}{C}}-OPO_3^{-2}$$

ATCase → PO_4^{-3}

Carbamoyl aspartate

d. Molecules synthesized in steps before the committed step in a metabolic pathway can have a number of metabolic fates. The products of a committed step are restricted to a single, specific purpose. The products of the committed step in pyrimidine synthesis are used only to produce pyrimidines.

e. CTP is an allosteric inhibitor, while ATP is an allosteric activator for ATCase.

9. a. **Ribonucleotide diphosphate + NADPH + H⁺ ------>**

   Deoxyribonucleotide diphosphate + NADP⁺ + H₂O

Ribonucleotide reductase is the enzyme, with thioredoxin and thioredoxin reductase playing an essential role in the shuttling of the electrons from NADPH to the dNDP molecule.

Chapter 17 The Metabolism of Nitrogen

b. This is a redox reaction in which NDP becomes reduced to dNDP, as NADPH becomes oxidized to NADP+.
c. dADP, dGDP, dCDP and dUDP.
d. dTMP is synthesized from dUMP by the enzyme, thymidylate synthetase, with N^5, N^{10}-methylenetetrahydrofolate serving as the methyl donor.

Biosynthesis of dTMP

10. The inhibition of hemoglobin biosynthesis can lead to anemia. An integral component of each subunit in hemoglobin is the heme cofactor, which is essential for the transport of oxygen to the cells. In the initial phase of heme biosynthesis, succinyl-

Chapter 17 The Metabolism of Nitrogen

CoA condenses with glycine, and following the loss of CO_2, forms δ-amino levulinate. The latter reaction is inhibited by Pb^{+2}. This leads to decreased levels of hemoglobin.

11. a. True.
    b. False. Amino acids are not stored. They are catabolized if they are not used for protein synthesis.
    c. True.
    d. False. Although pyridoxal phosphate is the coenzyme in the transamination reactions, the Schiff base linkage is to the amino acid and not the enzyme.
    e. True.
    f. True.
    g. False. It involves only one reaction and requires less energy than the *de novo* synthesis of purine nucleotides.
    h. True.
    i. False. Fluorouracil, although used in cancer chemotherapy, is an inhibitor for thymidylate synthase.
    j. True.

18

Metabolism in Perspective

An enormous number of diverse, yet interconnected reactions are occurring at any one time in the life of every cell. Because of this interdependence, these metabolic pathways must be exquisitely regulated within the cell. In addition, an ongoing intercellular communication network in multicellular organisms is essential to coordinate the short- and the long-term activities of the individual cells and tissues for homeostasis and for the optimum functioning of the organism. This regulation is carried out at many levels. Within the cell, metabolic pathways are regulated by allosteric enzymes in feedback loops, and to many processes are limited to certain organelles. Many metabolic intermediates, therefore, must be transported from one compartment to another. Furthermore, intercellular communications are carried out by **hormones**, which bind to receptors either on or within the target cells. A variety of mechanisms, some including secondary messengers, are then engaged to control the desired metabolic changes within the cell. The body of each person is sustained by the daily intake of **nutrients**, including macronutrients, micronutrients and minerals. The selection of these nutrients encompasses your diet, which can contribute to

Chapter 18 Metabolism in Perspective

promoting good health. The **immune response** is a collective protective action resulting from the two types of lymphocytes, the T cells and the B cells. Together the action of killer T, helper T, and memory T cells with antibody glycoproteins function to destroy pathogens damaging to the organism. The diatomic molecule, **nitric oxide**, exerts an enormous effect on many cell types. It is synthesized from arginine; the reaction is catalyzed by the activated enzyme, nitric oxide synthase. It plays a role in the immune response, the transmission of nerve impulses, and in the control of blood pressure. The effect of NO in cancer cells results from its inhibitory action in the citric acid cycle, the electron transport chain and in nucleotide biosynthesis.

Some Key Figures Revisited

Figure 18.1 (next page) A summary of catabolism, showing the central role of the citric acid cycle. Note tht the end-products of the catabolism of carbohydrates, lipids, and amino acids all appear. (PEP is phosphoenolpyruvate; α-KG is α-ketoglutarate; TA is transamination; ---->---->----> is a multistep pathway.)

Chapter 18 Metabolism in Perspective

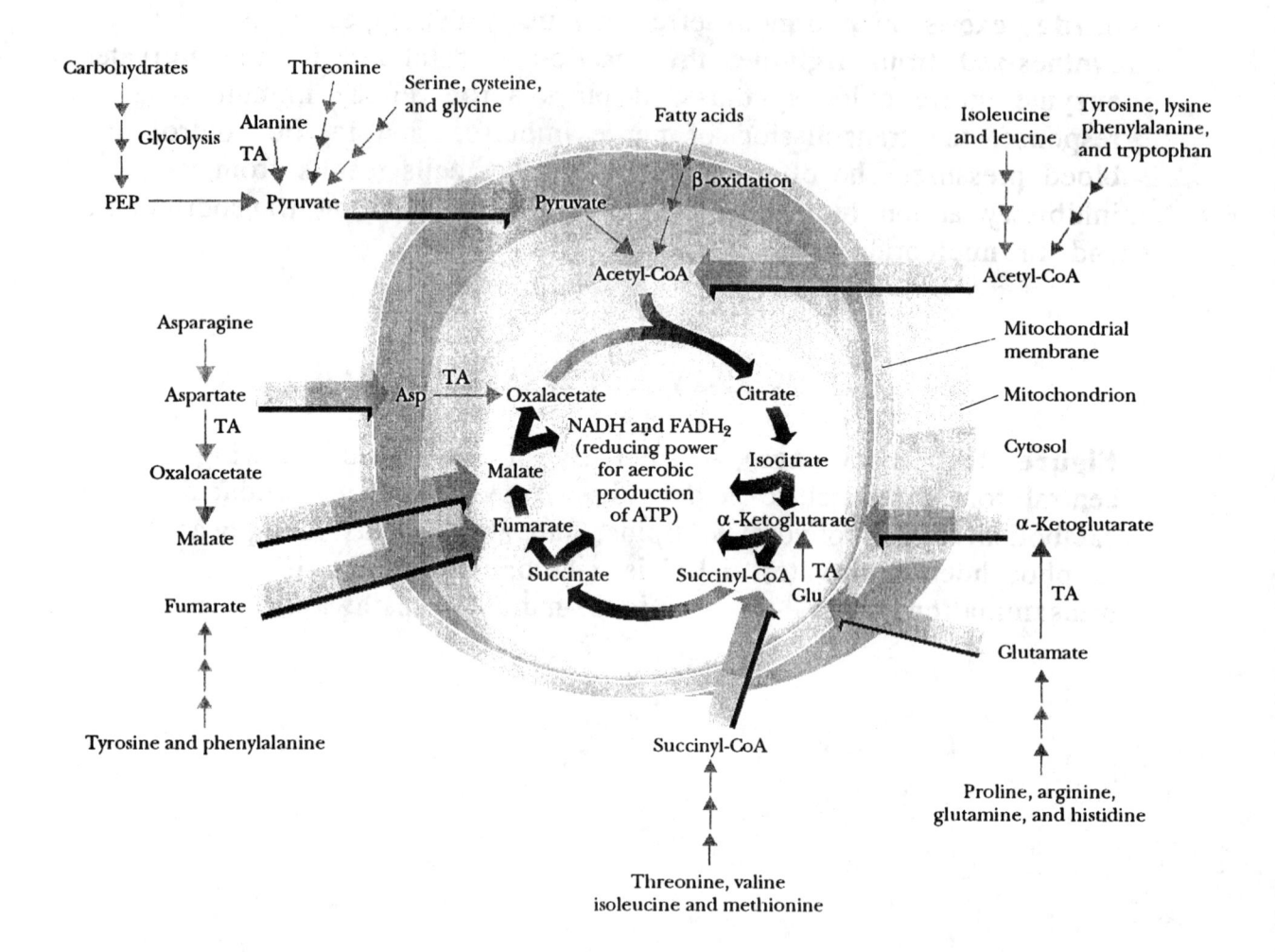

Chapter 18 Metabolism in Perspective

Figure 18.12 The PIP$_2$ second messenger scheme. When a hormone binds to a receptor, it activates phospholipase C, in a process mediated by a G protein. Phospholipase C hydroyzes PIP$_2$ to IP$_3$ and DAG. IP$_3$ stimulates the release of Ca^{2+} from intracellular reservoirs in the ER. A complex formed between Ca^{2+} and the calcium-binding protein calmodulin activates a cytosolic protein kinase for phosphorylation of a target enzyme. DAG remains bound to the plasma membrane, where it activates the membrane-bound protein kinase C (PKC). PKC is involved in the phosphorylation of a number of target enzymes. PKC also phosphorylates channel proteins that control the flow of Ca^{2+} in and out of the cell. Ca^{2+} from extracellular sources can produce sustained responses even when the supply of Ca^{2+} in intracellular reservoirs is exhausted.

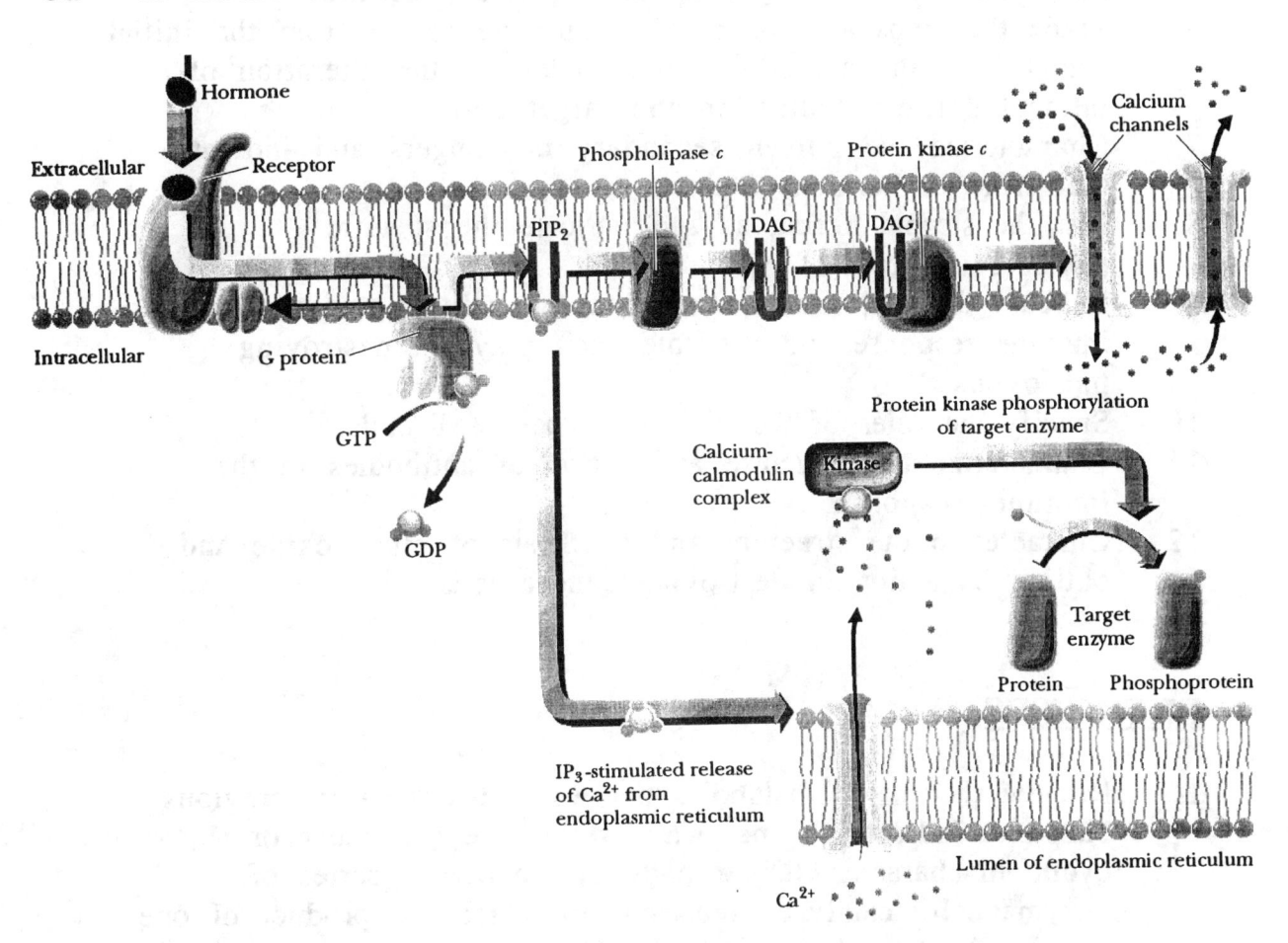

Chapter 18 Metabolism in Perspective

CHAPTER OBJECTIVES

1. Discuss the features of the citric acid cycle and describe its central role in cellular metabolism.
2. Make a list of the major metabolic pathways and indicate the cellular location where each takes place.
3. Catalog nutrients as macronutrients, micronutrients, and minerals and indicate their role in providing proper nutrition.
4. Indicate the key reactions and metabolic intermediates in the anabolism of carbohydrates.
5. Specify the role of malate dehydrogenase, malic enzyme and malate in the transport of oxaloacetate across the mitochondrial membrane for carbohydrate and lipid anabolism.
6. Trace the steps associated with hormone action from the initial signal from the central nervous system to the alteration of intracellular metabolism in the target cell.
7. Characterize the known secondary messengers and indicate their role in hormone action.
8. List the steroid hormones and indicate the specific physiological role of each.
9. Indicate the major components of the cellular and molecular immune response and the role each plays in destroying pathogens.
10. Specify the roles of the different types of T cells.
11. Characterize the structure and action of antibodies in the immune response.
12. Characterize the structure and synthesis of nitric oxide and outline its action in destroying cancer cells.

EXERCISES

1. Review the major metabolic pathways discussed in previous chapters. Indicate the pathways which are (a) linear or (b) cyclic in character. Linear pathways involve a series of enzymatically catalyzed reactions in which the product of one reaction is the substrate for the following enzyme. A cyclic pathway is similar, but requires that one of the metabolites be regenerated at the completion of the cycle.

Chapter 18 Metabolism in Perspective

2. Suggest a reason for the possible toxicity associated with the ingestion of high levels of fat-soluble vitamins.

3. Oxaloacetate is a metabolite of central importance in the life of the cell. It takes part in the citric acid cycle and is a key intermediate in gluconeogenesis and amino acid biosynthesis. The level of oxaloacetate in the cell must be regulated so that it can the accommodate acetyl-CoA that enters the citric acid cycle. In mammals, oxaloacetate can be synthesized in the mitochondria by the enzymatic carboxylation of pyruvate by pyruvate carboxylase (PC).

$$\text{pyruvate} + CO_2 + ATP + H_2O \xrightarrow{PC} \text{oxaloacetate} + ADP + P_i + 2 H^+$$

The activity of PC is regulated by the level of acetyl-CoA in the mitochondria. Suggest a general mechanism for this regulation.

4. Both hormones and neurotransmitters take part in intercellular communications. What property or properties distinguish a hormone from a neurotransmitter?

5. List the general functions of hormones and give an example of a hormone which fits into each category.

6. A human gene, referred to as the *ras* gene, exists in normal cells; however, specific mutations (changes in one or more of the bases in the DNA) can produce severe consequences to the organism. These mutations can cause the originally normal cell to undergo a transformation into a rapidly dividing tumor cell. In fact, it has been established that a number of mutated genes are associated with the occurrence of human cancers. Genes that have this characteristic are called cellular **"oncogenes"**. The normal *ras* gene codes for a 21 kDalton protein, in which there are regions homologous (very similar in sequence) to the sequence of the α-subunit of the G protein. Like the G protein, the normal protein product of the *ras* gene is a guanine nucleotide binding protein and it hydrolyzes GTP (exhibits a GTPase activity). However, although the protein product of the *ras* oncogene binds GTP, it has little or no GTPase activity.

Chapter 18 Metabolism in Perspective

    a. Suggest where you might expect to find the protein product of the *ras* oncogene in the cell.

    b. Suggest a possible mechanism of action for this protein, which might stimulate cell growth.

7. A group of natural products (naturally occurring organic compounds) are classified as phorbol esters. These natural products can stimulate the formation of tumors in animals when they are applied simultaneously with a carcinogen (a compound which produces cancer in animals). It has been established that phorbol esters produce this effect by activating protein kinase C.

    a. What are the names of the compounds with the abbreviations PIP_2, IP_3 and DAG?

    b. Which (PIP_2, IP_3, DAG) are considered to be second messengers?

    c. From an examination of the structures of the phorbol ester with IP_3 and DAG shown below, suggest a mechanism for the action of phorbol esters.

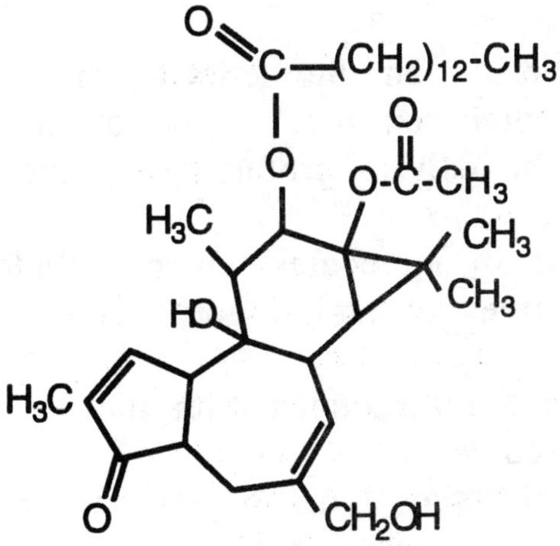

Chapter 18 Metabolism in Perspective

Phorbol ester

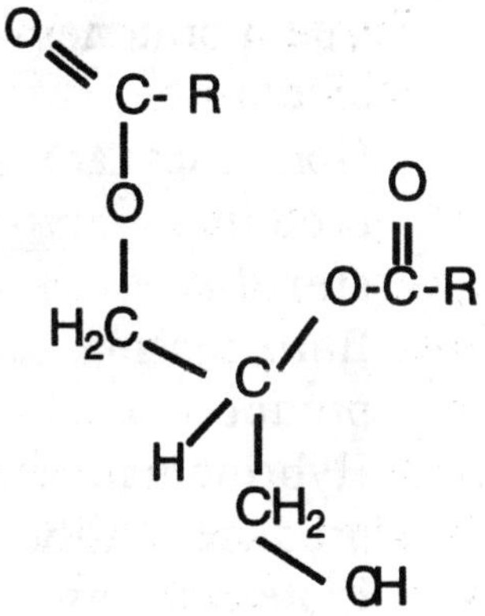

Diacylglycerol

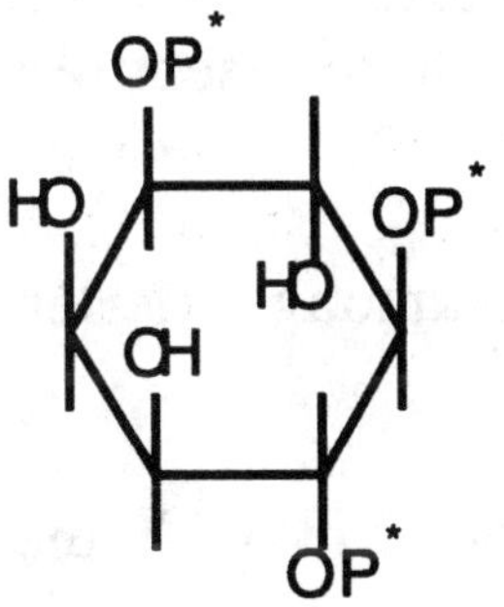

Inositol 1,4,5-triphosphate

$P^* = PO_3^{-2}$

8. Indicate whether the following statements are true or false. Correct the false statements.
   a. Oxaloacetate produced in the mitochondria can readily cross the membrane and enter the cytosol.
   b. The B vitamins are essential nutrients since they cannot be synthesized in humans and are the precursors for metabolically important coenzymes.
   c. The Food and Nutrition Board in the United States suggests a recommended daily allowance of nutrients for the average man or woman. For all the nutrients, the recommended amounts are the same or greater for men than for women.
   d. A reaction (such as the carboxylation of pyruvate to form oxaloacetate) that maintains an adequate level of metabolic intermediates is called an anaplerotic reaction.

Chapter 18 Metabolism in Perspective

    e. All hormones exert their effect by an interaction with a plasma membrane receptor protein on the outside of the cell. The hormones do not enter the cell to produce their metabolic effect.

    f. Hormones are a diverse group of molecules which include proteins, polypeptides, derivatives of amino acids and steroids.

    g. Both plasma cells and T cells are associated with the production of antibody molecules.

    h. Hybridomas are derived from fusing lymphocytes, which produce antibodies with normal mouse cells.

    i. Protein kinase C binds to adenylate cyclase to produce cAMP from ATP.

    j. The metabolic intermediates of the citric acid cycle, which can cross the mitochondrial membrane and leave the mitochondria are, citrate, isocitrate, succinyl-CoA and malate.

9. What is the relationship among interleukins, cytokines and lymphokines?

10. How does nitric oxide interfere with the cellular division in cancerous cells?

ANSWERS TO EXERCISES

1. Linear metabolic pathways include most of the pathways discussed, including glycolysis, gluconeogenesis, electron transport, steroid biosynthesis, etc., with the exception of those listed below as cyclic pathways.

   Cyclic pathways include the citric acid cycle, urea cycle and the Calvin cycle. The β-oxidation pathway for fatty acids can also be regarded as a cyclic pathway. (This pathway can be regarded as spiral in character in that the acyl chain is progressively shortened by C_2 units).

2. Fat soluble vitamins accumulate in the adipose tissue and because of their inherent solubility, can remain there for long

Chapter 18 Metabolism in Perspective

periods of time. They are not as readily excreted as are the water-soluble vitamins. Therefore, excessively high levels of fat soluble vitamins can accumulate in the fat to produce deleterious effects. On the other hand, excesses of water soluble vitamins (such as vitamin C) can be readily excreted.

3. Acetyl-CoA is a powerful allosteric effector of pyruvate carboxylase; the acetyl-CoA interacts with PC to stimulate enzyme activity. In fact, PC is virtually inactive when acetyl-CoA is not bound to it. At low acetyl-CoA levels, the acetyl-CoA enters the TCA cycle and combines with oxaloacetate to form citrate. It does not bind to PC. However, at high acetyl-CoA levels, it enters the citric acid cycle and also interacts with PC. This interaction stimulates the synthesis of oxaloacetate at sufficient levels to react with the intermediates derived from the acetyl-CoA entering the cycle. In addition, a fraction of the oxaloacetate can be diverted from the cycle in order to produce phosphoenolpyruvate and subsequently, glucose via gluconeogenesis.

4. Both hormones and neurotransmitters are synthesized in specific cells, and then secret and bind to receptors to produce or alter a specific biological activity. In fact, a number of molecules epinephrine (adrenaline) and norepinephrine have been characterized as both hormones and neurotransmitters. Therefore, the distinction cannot be assigned simply on the chemical nature of the molecule. The definition must rely on the function it carries out (i.e., its physiological nature). A major distinction between hormones and neurotransmitters depends on the distance over which it acts. Neurotransmitters act over the very short distance across a synapse (2×10^{-6} cm), while a hormones act over long distances (many centimeters). Hormones are secreted from an endocrine gland and travel through the blood stream to interact with a distant target cell and to stimulate metabolic activities.

5.
	Function	**Example**
1.	Homeostasis	Insulin and glucagon (in controlling glucose levels)
2.	Mediate a response to external stimuli	Epinephrine and norepinephrine

Chapter 18 Metabolism in Perspective

3. Growth/development   Female or male sex hormones

6. a. The *ras* protein would be expected to be a component of the plasma membrane similar to that of the G proteins. This is, in fact, where it is actually found in the cell.
   b. It most likely exhibits the same function as that of the G proteins. That is, it is involved in signal transduction, which results in the production of cAMP as a result of hormone binding to the plasma membrane receptor. Recall that the G protein is in the inactive form when GDP is bound to the α-subunit and becomes activated when GTP binds it. Upon the hydrolysis of GTP to GDP, the protein again is cycled back to the inactive form. However, since the oncogenic *ras* protein does not have a GTPase activity, it is persistently in the activated form and, therefore, raises the level of cellular metabolism. This continuously increased activity would disrupt the regulation of normal growth and possibly produce unregulated cell division. Although this theory remains to be proven, it is a major avenue of cancer research currently being pursued by many biomedical researchers in the world.

7. a. PIP$_2$;   Phosphatidylinositol 4,5-bisphosphate
      IP$_3$;   Inositol 1,4,5-triphosphate
      DAG;   Diacylglycerol (This must be a **1,2**-diacylglycerol.)
   b. The second messengers are IP$_3$ and DAG.
   c. The structure of DAG and the upper segment of the phorbol ester are very similar. This similarity suggests that these molecules may exhibit comparable solubilities in the lipid bilayer, interact with protein kinase C by similar interactions and consequently act by similar mechanisms in activating protein kinase C.

8.
   a. False. Oxaloacetate must be transported by the malate transport mechanism.
   b. True.
   c. False. (Refer to Table 18.1 in the text). This is true for all nutrients except Fe, for which a higher level is recommended for women.

Chapter 18 Metabolism in Perspective

    d.   True.

    e.   False. Not all hormones act in this manner. Steroid hormones enter the cell and bind to receptors within the cell.

    f.   True.

    g.   False. Both B cells and plasma cells are associated with the production of antibody molecules.

    h.   False. Hybridomas are derived from the fusion of lymphocytes, which produce antibodies, and mouse myeloma cells. The mouse myeloma cells are immortal and will grow c continuously. Therefore, each hybridoma cell produces one specific antibody (monoclonal) and grows continuously. This growth provides a ready route to obtain essentially an unlimited supply of the specific antibody.

    i.   False. It phosphorylates target proteins and enzymes, including those proteins that regulate the flow of Ca^{+2} ions into and out of the cell.

    j.   True.

9. Cytokines are soluble proteins produced by one cell while they act on another cell. Lymphokines are cytokines which play a specific role with lymphocytes. Interleukins are cytokines produced by macrophages when bound to T cells. The terms used for these soluble communication proteins, which are associated with the immune response, range from a general classification to a more specific one.

10. First, it inhibits the enzyme, aconitase, which is essential in the early steps of the citric acid cycle. This is the "hub" of metabolic activity for the cell. Second, it inhibits Complex I in the electron transport chain, resulting in the shut down of ATP production, which is essential for a multitude of activities in the cell, including anabolic pathways. Nitric oxide also inhibits ribonucleotide reductase, which is required for the synthesis of deoxyribonucleotides and DNA. All cells, especially those cells that are rapidly dividing, are effected by this multipronged attack by NO.

19

Biosynthesis of Nucleic Acids: Replication and Transcription of the Genetic Code

DNA contains the genetic information in the form of a sequence of nucleotides. By the process of DNA **replication**, this information is passed on from one generation to the next with a remarkable fidelity. The DNA is synthesized in a continuous manner on the **leading strand** and in a discontinuous manner on the lagging strand of the **replication fork**. In *E. coli*, this is accomplished by a multitude of proteins and enzymes, including gyrase, helicase, single-strand DNA binding proteins, primase, DNA ligase and DNA polymerase III and I. Both of these DNA polymerases exhibit not only a polymerization activity, but also exonuclease activities to help attain a high fidelity in DNA replication. The importance of DNA is evident by the finding that it is the only molecule in the cell that can

Chapter 19 Biosynthesis of Nucleic Acids

be enzymatically repaired if it is chemically or physically modified. The **cut-and-patch repair mechanisms** for DNA alterations during replication and on existing DNA are discussed. **Transcription** is the process of expressing the genetic material into RNA. All three forms of RNA undergo some degree of post-transcriptional processing; however, the most elaborate processing occurs on immature mRNA in eukaryotic cells. Both the 5' and 3' ends of mRNA become modified, the intron segments of the RNA are excised out and the exon segments spliced together. A number of RNA molecules have been discovered, which exhibit catalytic ability, and are classified as either Group I or Group II **ribozymes**.

CHAPTER OBJECTIVES

1. Outline the role of nucleic acids in replication, transcription and translation.
2. Discuss the experimental support for a semi-conservative mechanism of replication.
3. Summarize the structure and the functional characteristics for *E. coli* DNA polymerase I, II and III.
4. Identify the enzymatic activities of DNA polymerase I and III and indicate their role in DNA replication and DNA repair.
5. List the proteins and enzymes associated with replication at the replication fork and indicate the function of each.
6. Indicate how DNA replication differs in prokaryotes and eukaryotes.
7. Characterize the cut-and-patch process in the excision repair pathway that corrects chemically or physically altered segments of DNA.
8. List the ways in which spontaneous and induced mutations of DNA can occur in a living cell.
9. Outline the characteristics of *E. coli* RNA polymerase and describe the steps in the synthesis of RNA.
10. Describe the post-transcriptional processing that takes place on immature mRNA in the nucleus of the cell.
11. Outline the known processes in which RNA serves as an enzyme and briefly outline the mechanisms of action.
12. Define the term, ribozyme, indicate the molecules upon which they act, and describe the general mechanisms of action for a Group I and a Group II ribozyme.

Chapter 19 Biosynthesis of Nucleic Acids

SPOTLIGHT

RNA as an Enzyme: The Changing of a Mind-Set

Transfer RNA (t-RNA) molecules play an essential role in the process of protein synthesis. Each t-RNA complexes with a specific amino acid, thereby activating it to the form utilized in protein synthesis.

t-RNA + amino acid ---> [t-RNA-amino acid]

However, prior to this step, a number of processing events must occur with high fidelity. The form of the t-RNA molecule produced initially in the nucleus (referred to as immature t-RNA), has additional segments of nucleotides on both its 5' and the 3' ends (solid black lines in the Figure below). These segments must be cleaved to yield the mature t-RNA (shown as the thicker bands in the Figure).

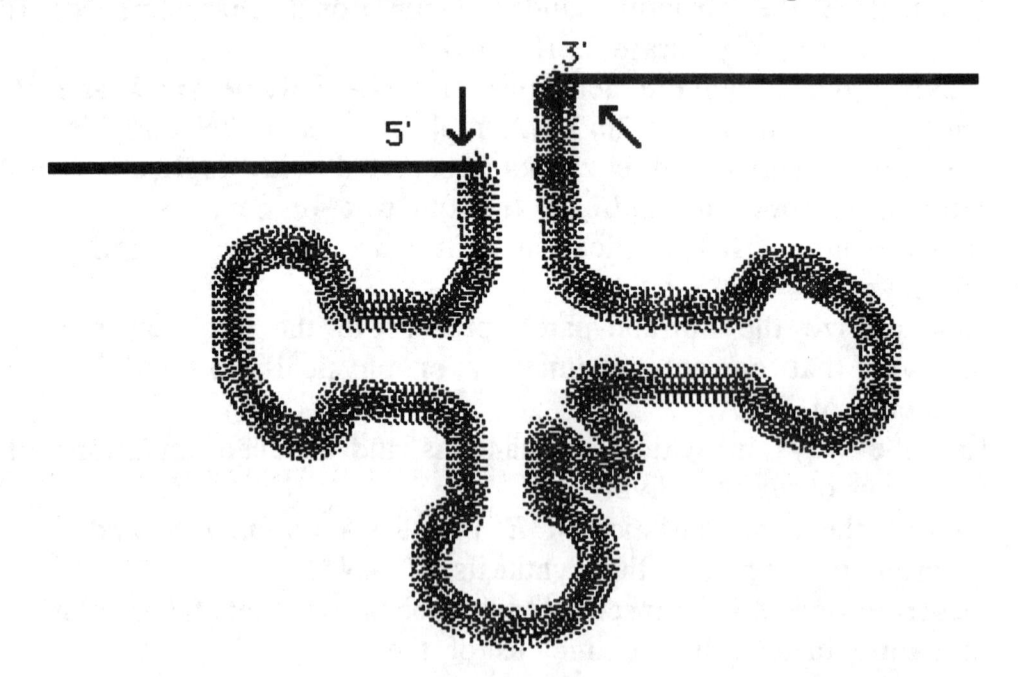

It had long been known that the enzyme, which cleaves the 5' end of the immature t-RNA, was the endonuclease, RNase P. This enzyme complex is composed of both a 14 kDalton protein and a 377-nucleotide RNA molecule. This remarkable enzyme correctly cleaves not only all 60 or so

Chapter 19 Biosynthesis of Nucleic Acids

different t-RNAs in *E. coli*, but it can also carry out this processing event on precursor t-RNAs from other organisms. It was assumed that the protein was the catalyst, with the RNA possibly conferring stability and substrate specificity in the enzyme. Attempts to define the role of each subunit revealed that in physiological solution, containing 5-10 mM Mg^{+2}, the protein alone had no hydrolytic activity. Subsequent experiments examined the RNA and protein subunits separately and in all combinations for catalytic activity. The protein again exhibited no activity under any conditions. To the astonishment of Sidney Altman and his associates of Yale University, the RNA alone showed activity in solution conditions in which the Mg^{+2} concentration was higher than normally used. To insure that there was no trace amounts of protein in the RNA sample, recombinant DNA techniques were used to prepare the pure RNA subunit of RNase P. This experiment showed this same activity and proved beyond any doubt that the RNA molecule exhibited the catalytic activity. The higher Mg^{+2} concentration that was required for activity of the lone RNA subunit suggested, and later it was shown to be the case, that the protein subunit in RNase P served in the complete enzyme (holoenzyme) as a "structural counter-ion" to provide stability. The enzyme obeys Michaelis-Menten kinetics with a K_M value of 5×10^{-7} M.

At this same time, Thomas Cech and his colleagues at the University of Colorado were studying the rRNA genes in the single cell eukaryote, *Tetrahymena thermophila*, and they were intrigued by the process in which the intervening sequences (introns) were excised in the processing of the pre-rRNA. They found that when the isolated pre-rRNA (containing the introns) was incubated with guanosine, in the absence of any protein, the intron excised itself and spliced together the flanking exons. This and subsequent studies showed that this reaction was autocatalytic, with the 413 nucleotide RNA acting as both the substrate and the catalyst, with a K_M value of 32 micromolar. A few other RNA molecules have since been shown to also exhibit catalytic activity.

These findings suggest that RNA molecules may have been the original catalysts, but that they were restricted in the types of possible reactions they could catalyze due to the

Chapter 19 Biosynthesis of Nucleic Acids

limited number of nucleotides in RNA. It is further speculated that during evolution, protein molecules, with twenty different amino acid residues to draw on for a very wide range of structures and activities, became the more versatile and numerous biological catalysts.

Sidney Altman and Thomas Cech were awarded the Nobel Prize in Chemistry in 1989 for this seminal work which expanded the concept of enzymes and catalytic activity in biological molecules.

EXERCISES

1. Describe two biological processes that can produce spontaneous mutations in DNA.

2. Name three exogenous agents (physical or chemical) that can induce mutations in DNA.

3.
   a. What is a snRNP?
   b. What is the composition and role of a spliceosome?
   c. In what types of cell is the spliceosome found and where is their location within the cell?

4. Draw a double-stranded DNA fragment that can be used as a template and primer for DNA synthesis by DNA polymerase I. Indicate the role of each strand in DNA synthesis.

5. Arthur Kornberg isolated and characterized the first DNA polymerase from *E. coli* bacteria. A few years after this work on DNA polymerase I, a mutant form of *E. coli* was discovered that had the following characteristics:
   a. The mutant bacteria had only 1% of the DNA polymerase I activity found in wild type *E. coli* cells; however, cell division occurred at the same rate as in normal *E. coli* cells.
   b. The mutant *E. coli* cells were very sensitive to exposure to UV light, with a high percentage of the cells dying after irradiation.

What do these findings suggest about the possible function of DNA polymerase I in *E. coli* cells?

Chapter 19 Biosynthesis of Nucleic Acids

6. Meselson and Stahl carried out a now classic density-gradient experiment in which they showed that DNA replication occurred by a semi-conservative mechanism. Could the incorporation of ^{32}P and ^{31}P into the cells have worked equally well in these experiments as a replacement for ^{14}N and ^{15}N labeling? Explain briefly.

7.
   a. What precursor molecules are needed for DNA synthesis?
   b. In the formation of the phosphodiester backbone in DNA, what group on the DNA primer strand serves as the nucleophilic attacking group? What electrophilic group is the site of the attack? What are the products in this reaction?
   c. Is a high energy bond produced in the synthesis of DNA?

8. Indicate the enzymatic activities associated with *E. coli* DNA polymerase I, II and III.

9. Indicate the known cellular functions of *E. coli* DNA polymerase I, II and III.

10. Indicate the role of the following proteins in DNA replication in *E. coli*.
    a. Gyrase
    b. DNA single-strand binding (SSB) proteins
    c. Helicase
    d. Primase
    e. DNA polymerase I
    f. DNA polymerase II
    g. DNA polymerase III

11. Consider the excision repair of a lesion in existing DNA. Envision a logical stepwise process for eliminating this lesion and replacing the altered DNA segment with the correct DNA segment. What types of enzymes would participate in the excision repair process in *E. coli* ?

12. DNA polymerase I serves primarily a repair role in *E. coli*, while polymerase III is the enzyme primarily responsible for DNA synthesis. What are the corresponding enzymes that carry out these functions in eukaryotes?

Chapter 19 Biosynthesis of Nucleic Acids

13. What is the difference in composition between the core and the holoenzyme for *E. coli* RNA polymerase? How does this difference relate to the functional roles of these two RNA polymerase complexes?

14. Indicate the three types of post-transcriptional modifications that are carried out on immature tRNA molecules.

15. Compare the structural gene, the corresponding immature mRNA transcript synthesized directly by RNA polymerase, and the mature mRNA transcript, which leaves the nucleus for translation in the cytoplasm of eukaryotes.

16. Indicate whether the following statements are true or false. If the statement is false, correct it.
    a. All DNA polymerases synthesize DNA in a 5'---> 3' direction.
    b. DNA synthesis requires a 5'-hydroxyl group on the primer strand.
    c. Every "replication bubble" contains two replication forks.
    d. Okazaki fragments are the RNA fragments in intron units.
    e. Discontinuous DNA synthesis occurs on the leading strand of DNA.
    f. *E. coli* DNA polymerase I is a more abundant protein in *E. coli* cells than is DNA polymerase III.
    g. Reverse transcriptase is a (RNA directed) DNA polymerase, which uses RNA as the template.
    h. DNA polymerase β in eukaryotic cells is thought to be a repair enzyme.
    i. Group II ribozymes have a requirement for a free guanosine for activity.
    j. The rho protein is involved in the initiation of transcription by *E. coli* RNA polymerase.

ANSWERS TO EXERCISES

1. Mutations can occur spontaneously during the process of replication and in the deamination of cytosine.

Chapter 19 Biosynthesis of Nucleic Acids

2. Three exogenous agents that can cause mutations are UV radiation, X-rays (both physical agents), and a large number of chemicals (mutagens).

3. a. A snRNP is a complex containing both RNA and protein, which exhibits enzymatic activity essential in the splicing process.
    b. The spliceosome is a larger assemblage, containing the snRNP and other factors, and is the unit responsible for the actual splicing.
    c. Spliceosomes are found in the nucleus of eukaryotic cells. They are not found in prokaryotic cells.

4. The primer strand is the strand on which DNA synthesis occurs. It must have a terminal 3'-hydroxyl deoxyribose for synthesis (to initiate or continue). The template strand is complementary to the primer strand. Therefore, the base on the template strand, because of the demand for specific base pairing, determines the nucleotide that can be inserted in the primer strand.

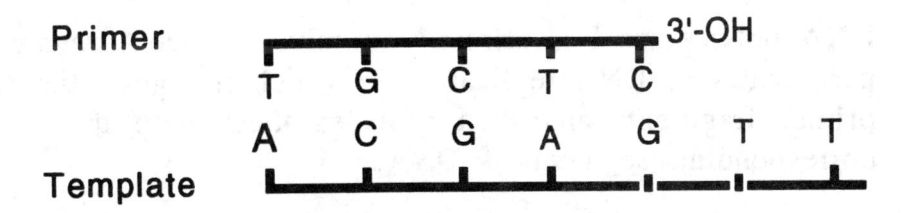

```
Primer                               3'-OH
         T    G    C    T    C
         A    C    G    A    G    T    T
Template
```

5. The first observation suggested that DNA polymerase I may not be the primary polymerase involved in DNA replication. The second observation suggested that DNA polymerase I might play a prominent role in the repair of UV damaged DNA.

6. The use of isotopes, ^{32}P and ^{31}P, would be more difficult than the ^{14}N and ^{15}N isotopes used in these experiments. The ^{14}N and ^{15}N isotopes are stable and the percent difference in mass is greater here than in the phosphorus isotopes. Although the ^{31}P isotope is stable, the ^{32}P isotope is radioactive and will decay with a half-life of about 15 days. In addition, there are more nitrogen atoms in DNA than there are phosphorus atoms.

Chapter 19 Biosynthesis of Nucleic Acids

7. a. DNA synthesis (i.e., DNA synthesis *in vitro*, or in a test tube) is a far simpler process than DNA replication (i.e., DNA synthesis *in vivo* or in a living organism). DNA synthesis requires a DNA primer (with a 3'-OH on the deoxyribose), a DNA template strand, Mg^{+2} ion, the four deoxyribonucleoside triphosphates, and DNA polymerase.
   b. The 3'-OH in the deoxyribose is the nucleophile that attacks the electrophilic phosphate group on the dNTP. The phosphate group is the one directly attached to the deoxyribose in the incoming nucleoside triphosphate (the α-phosphate in the nucleoside triphosphate). The products of the reaction are DNA and pyrophosphate (P_iP_i).
   c. No. The phosphate ester bond is a low energy bond.

8. All three DNA polymerases from *E. coli* exhibit the 5' --> 3' polymerization and the 3' --> 5' exonuclease activities. DNA polymerase I and III also have the 5' --> 3' exonuclease activity, while polymerase II does not have this enzymatic capability.

9. DNA polymerase I is primarily involved in DNA repair. It also participates in DNA replication in which it digests the RNA primer fragments and then replaces them with the corresponding segment of DNA.

   The function of DNA polymerase II is not known.

   DNA polymerase III is the polymerase primarily responsible for DNA replication in *E. coli*.

10. a. Gyrase removes supercoils in front of the replication fork.
    b. Single-strand binding proteins bind **preferentially** to single-stranded DNA. This stabilizes them so that replication can occur.
    c. Helicase promotes unwinding of the DNA at the replication fork.
    d. Primase synthesizes a small fragment of RNA which serves as a primer for DNA synthesis.
    e. DNA polymerase I digests the RNA primer and synthesizes a DNA segment to fill the gap between the Okazaki fragments.

Chapter 19 Biosynthesis of Nucleic Acids

    f. DNA polymerase II has no known function in replication.
    g. DNA polymerase III synthesizes most of the DNA in the replication process.

11. The types of enzymes that would participate in the excision repair process in *E. coli* are:
    a. An endonuclease that nicks the phosphodiester backbone on both sides of the damaged region.
    b. DNA polymerase I aids in the release of this damaged segment and synthesizes the correct DNA sequence (complementary to the undamaged strand) to fill in this gap.
    c. A DNA ligase joins the newly synthesized segment to the DNA.

12. Eukaryotic DNA polymerase α is the primary polymerase in DNA replication, while DNA polymerase β, like DNA polymerase I in *E. coli*, is primarily involved in DNA repair.

13. The core RNA polymerase contains four subunits, $\alpha_2\beta\beta'$, and carries out chain elongation of the RNA in transcription. The RNA polymerase holoenzyme, $\alpha_2\beta\beta'\sigma$, contains five subunits. The σ subunit is required for initiation of RNA synthesis at the correct nucleotide in DNA. The holoenzyme accomplishes this by binding to the promoter region of the gene. After initiation and during the elongation process, the σ subunit dissociates from the holoenzyme.

14. Immature tRNA molecules undergo trimming, addition of terminal sequences and the chemical modification of some of the bases.

15. The structural gene is a segment of double-stranded DNA. The immature mRNA is an single strand of RNA complementary to one of the strands of DNA. The immature transcript undergoes a rather extensive post-transcriptional processing. This involves the capping of the 5' end of the transcript with a unique trinucleotide. A guanylate residue, which is methylated at the N-7 position, is added to the 5' end. This is attached to the adjacent ribose by a 5'-5' triphosphate linkage. Intron segments of the RNA are excised and the exons are spliced

Chapter 19 Biosynthesis of Nucleic Acids

together. A polyadenylate "tail" is attached to the 3' end of the RNA to finish the mature RNA transcript. After this processing is completed in the nucleus, the mature transcript is transported into the cytoplasm for translation.

16. a. True.
    b. False. DNA synthesis requires a 3'-hydroxyl group on the primer strand.
    c. True.
    d. False. Okazaki fragments are the small DNA fragments which are synthesized discontinuously on the lagging strand of the replication fork.
    e. False. Discontinuous DNA synthesis occurs on the lagging strand of the replication fork.
    f. False. DNA polymerase III is more abundant than DNA polymerase I in *E. coli.*.
    g. True.
    h. True.
    i. False. Group I ribozymes require a free guanosine for activity. This is not a requirement for the Group II ribozymes.
    j. False. The rho protein is involved in the termination of the RNA transcript in *E. coli*.

20

Protein Synthesis: Translation of the Genetic Message

Proteins are the "worker-bees" of the cell. Proteins and enzymes are essential to cellular metabolic activity and to the integrity of the cell itself. The biosynthesis of proteins is continuously taking place in all living cells and is one of the most complex of all cellular processes, with the participation of over 300 macromolecules. This intricate process begins with the activation of the amino acids to be incorporated into the protein. After the initiation of the protein chain, at what will become the amino end of the protein, individual amino acids are added to form peptide bonds in the growing peptide chain. Termination of protein synthesis results in the release of the protein from the ribosome and the dissociation of the entire protein synthesizing complex, including the ribosomes, mRNA, tRNAs and protein factors. The genetic message in the mRNA is translated into the sequence of amino acid residues in the protein according to the **genetic code**. These trinucleotide

Chapter 20   Protein Synthesis

sequences in the mRNA form complementary hydrogen bonds with the anticodon triplets in the activated tRNA molecules. The primary regulation of gene expression in prokaryotes occurs at the level of **transcription** and is generally described by the **operon theory**. The lactose operon is a paradigm for the understanding of the expression of other genes in prokaryotic organisms. Cancer producing viruses (oncogenic viruses) such as the DNA virus, SV40 virus, and the RNA retroviruses, HTLV-I and HTLV-II, are introduced. Some of the prominent characteristics of the retrovirus which causes AIDS, the human immunodeficiency virus (HIV), are presented.

A Key Figure Revisited

Figure 20.13 (next page) The *lac* operon in the absence and presence of inducer. (a) When no inducer is present, the regulatory gene (I) is transcribed, leading to the production of repressor. The repressor in turn bonds to the operator gene (O), blocking transcription of the structure genes that code for ß-galactosidase and the other proteins produced under the control of the *lac* operon. (b) In the presence of inducer, the repressor is still produced but becomes inactive when the inducer is bound to it. The inactive repressor does not bind to the operator, and the structural genes of the *lac* operon are expressed. Other regions of the DNA shown here are the promoter (P) and the genes for the ß-galactosidase (Z), permease (Y), and acetyl transferase (A).

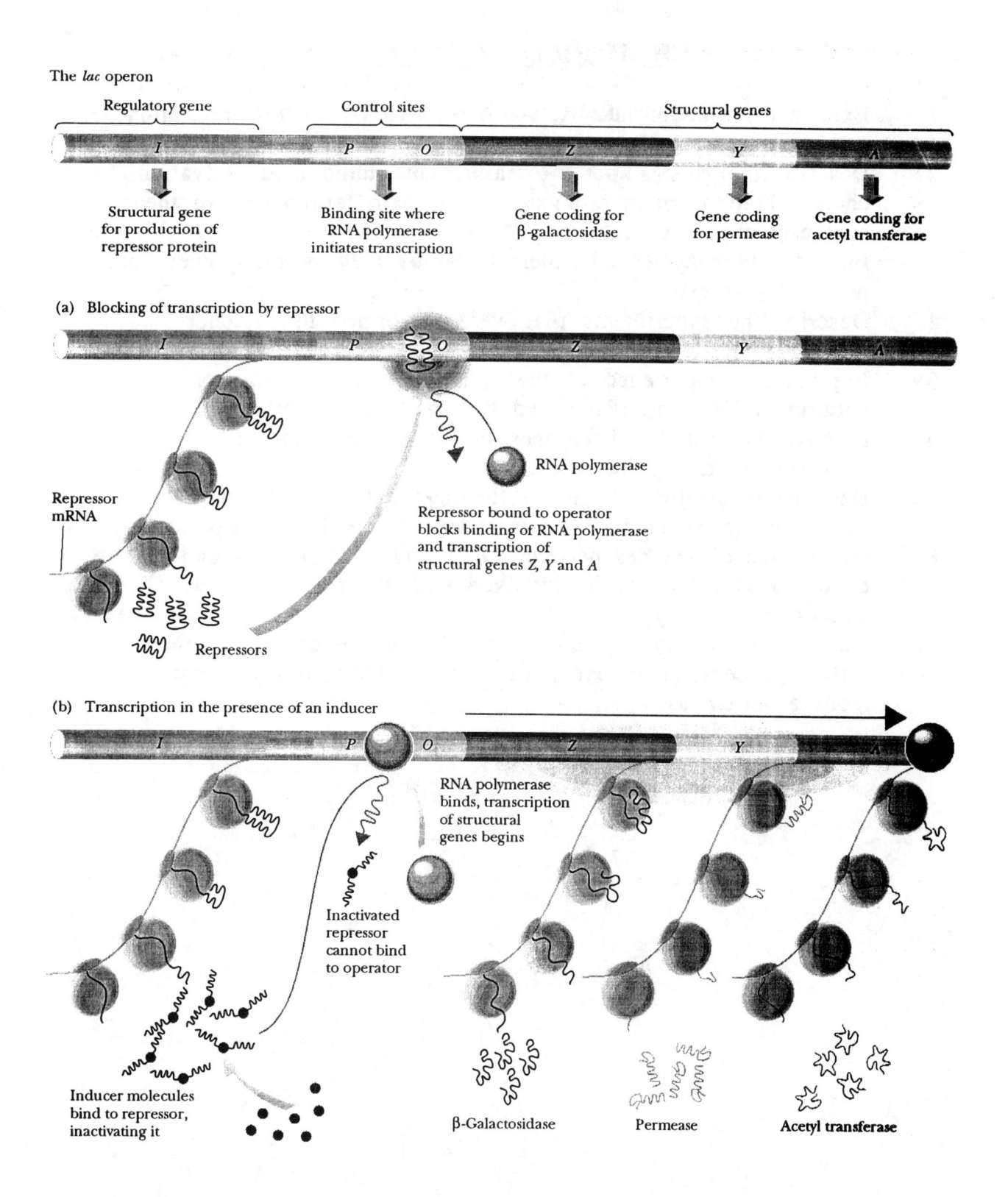

Chapter 20 Protein Synthesis

The *lac* operon

Regulatory gene	Control sites	Structural genes
I	P O	Z Y A

- I → Structural gene for production of repressor protein
- P → Binding site where RNA polymerase initiates transcription
- Z → Gene coding for β-galactosidase
- Y → Gene coding for permease
- A → Gene coding for acetyl transferase

(a) Blocking of transcription by repressor

Repressor mRNA

RNA polymerase

Repressor bound to operator blocks binding of RNA polymerase and transcription of structural genes Z, Y and A

Repressors

(b) Transcription in the presence of an inducer

RNA polymerase binds, transcription of structural genes begins

Inactivated repressor cannot bind to operator

Inducer molecules bind to repressor, inactivating it

β-Galactosidase   Permease   Acetyl transferase

265

Chapter 20 Protein Synthesis

CHAPTER OBJECTIVES

1. Explain the roles of mRNA, tRNAs and rRNAs in the translation process.
2. Describe the most important features of amino acid activation, chain initiation, chain elongation and chain termination in the synthesis of a protein.
3. Indicate the enzymes and their roles involved in the process of protein synthesis.
4. Describe the experiments that were performed to decipher the genetic code.
5. Explain the importance of the codon-anticodon interaction between mRNA and tRNA and the "wobble" hypothesis.
6. Discuss the general differences between translation in prokaryotes and eukaryotes.
7. Describe the major features of the operon theory of gene expression in prokaryotes as it relates to the lactose operon.
8. Outline the differences between DNA and RNA viruses and explain how they may be involved with the production of cancers.
9. Define the term oncogene and indicate an important general difference between an oncogene from a DNA virus and that from a retrovirus.

Chapter 20 Protein Synthesis

SPOTLIGHT

Cystic Fibrosis: The Gene, The Protein and A Present Day Candidate For Clinical Gene Therapy Trials

Cystic fibrosis (CF) is the most common lethal congenital disease in Caucasians. While one in every 20 individuals is a carrier, the incidence is 1 in 2,000 live births, with estimates that about 5,000 individuals die each year from CF. CF is a disease of the secretory epithelia. The cells in this tissue mediate transport of primarily water and salt from the blood to the outside world (i.e., the lumen). The clinical manifestations observed in afflicted individuals are the decrease in the secretion of fluid and the accumulation of a thick mucus in the bronchial tubes. This blocks the airways, promotes bacterial infections and after years of chronic lung infections, leads to heart and lung failure.

In 1989 the gene associated with CF was cloned! It is located on chromosome 7, spans 250 kilobases of DNA and has 24 exons. A comparison of the gene from normal individuals with those from a number of CF patients revealed that about 70% of the CF patients had <u>three adjacent bases missing</u> in the gene! (only 3 bases of the total 250,000 bases). This leads to 3 bases missing in the mRNA and, since this corresponds to a codon - the specific codon for phenylalanine - the CF protein had this <u>single amino acid residue deleted</u> at position 508 in this protein of 1,480 amino acid residues. A number of pieces of data, including the structure, clearly indicated to membrane biologists that it was a membrane protein. It was called the Cystic Fibrosis Transmembrane Regulator (CFTR). The CF gene is only expressed (i.e., transcribed and translated into protein) in certain cells of the pancreas, the sweat glands, and the lining of the respiratory tract. Consistent with this gene being associated with CF, these are the tissues in which the disease manifests itself. Previous findings had also indicated that the disease was not associated with the <u>absence</u> of chloride

Chapter 20 Protein Synthesis

ion channels, but with the <u>abnormal regulation</u> of the channels.

The figure below shows a tentative model for the CFTR in association with the membrane. The protein passes through the membrane (12 times). The amino- and carboxyl-ends of the protein are located in the cytoplasm, together with a region of the protein that has binding domains for ATP. This suggests that the protein acts as an ion pump, rather than as a simple ion channel. Note that the mutation is near the ATP binding domains.

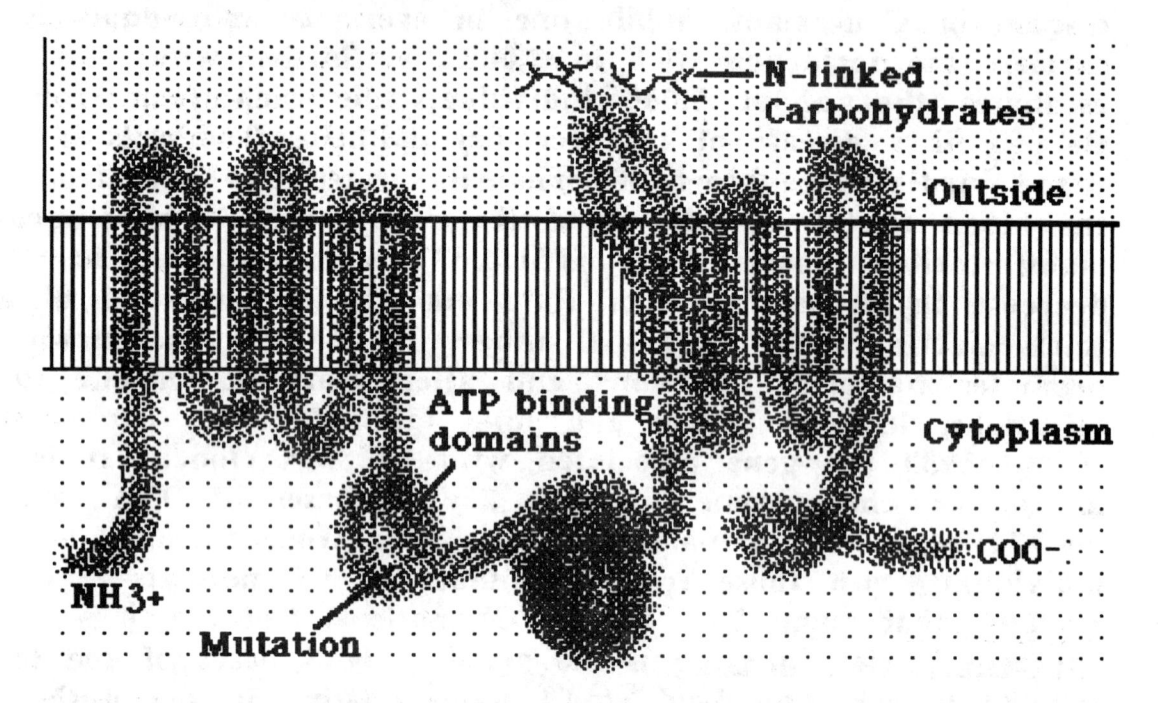

Knowing the molecular culprit that is responsible for the disease is a giant step in understanding CF, but it can be very far from providing an effective treatment for the disease. However, in September of 1990, stunning experiments were successfully performed on cells in culture which showed the following:

1. The <u>normal</u> CF gene was inserted into <u>cells</u> with a defective CF gene and therefore faulty CFTRs (clogged chloride ion channels). After the cells took up the gene, normal CFTRs were produced and the chloride ion channels were opened. This experiment showed that insertion of a normal gene into a living cell could be expressed to produce a physiologically significant change in the cell.

Chapter 20 Protein Synthesis

    2. This experimental finding also provided indisputable evidence that the isolated gene is, in fact, the cystic fibrosis gene and that its protein product is a chloride ion channel protein.

The future for finding a cure for CF is exciting; a number of recent developments look very promising. First, it has been known for sometime that much of the thick mucus in the lungs originates from the DNA from lysed bacteria. This mucus greatly limits the effectiveness of antibiotics, which are administered to stop a bacterial infection. An aerosol throat spray is now available, which contains DNase I, an enzyme which degrades DNA. The CF patient can use this spray to degrade the bacterial DNA, which significantly reduces the mucus and thereby permits the antibiotic to work more efficiently. Second, in the summer of 1994, scientists showed that they could successfully introduce a normal CF gene into the lungs of CF patients. This procedure was accomplished by using recombinant DNA technology with a modified cold virus, called adenovirus, as the vector. In these initial trials, the new gene lasted for only 10 days. The next step is to determine if repeated transfers of the normal gene into the lungs can be accomplished.

EXERCISES

1   a.   Write the overall reaction for the activation of an amino acid.
      b.   What enzyme catalyzes this reaction?
      c.   How many high energy bonds are hydrolyzed in the activation process?

2.   a.   Which end of the tRNA molecule binds to the amino acid?
      b.   Specify the exact site at which the amino acid is covalently bound to the tRNA molecule and the type of chemical linkage produced by the reaction.

3.   a.   What are the sizes of the subunits (in Svedberg units, S values) that make up the ribosome in *E. coli*? What is the size of the complete ribosome?

Chapter 20 Protein Synthesis

    b.   Describe the two types of interactions between the mRNA and the 16S rRNA in the small ribosomal subunit that determine where protein synthesis starts.

    c.   Which ends of the mRNA and the 16S rRNA molecules serve as starting points for the translation process?

    d.   Describe the composition of the Shine-Dalgarno sequence. Which RNA does this sequence reside on and where is its relative location in the RNA?

    e.   What is the DNA sequence that corresponds to the Shine-Dalgarno sequence?

4.   a.   DNA is synthesized in what direction?
      b.   DNA is read by DNA polymerase in what direction?
      c.   RNA is synthesized in what direction?
      d.   DNA is read by RNA polymerase in what direction?
      e.   Protein synthesis begins at what end of the mRNA molecule?
      f.   What is the orientation of the anticodon triplet in tRNA relative to the codon sequence on mRNA?
      g.   Which end of the protein is synthesized first?

5.   In prokaryotic organisms, translation occurs before the mRNA is complete, but this does not occur in eukaryotic cells. Explain briefly.

6.   a.   Define the term genetic code.
      b.   Synthetic RNA polynucleotides were used to help define the nucleotide triplets on mRNA which correspond to the amino acids which are incorporated into proteins. What proteins are produced by the translation of the following RNA molecules?
          i.   poly(UUU)
          ii.   poly(GCA)
          iii.   poly(AU)

7.   a.   What is the relative orientation of the codon and anticodon triplets as they base pair with each other?
      b.   What is the "wobble" hypothesis?

8.   The first 15 nucleotides in the DNA sequence of a gene are GGACCAGTCACACAT. This is the DNA strand that is read by

Chapter 20 Protein Synthesis

RNA polymerase. What is the corresponding amino acid sequence?

9. The protein that is coded for by the gene associated with the disease cystic fibrosis is 155.4 kDalton in size (considering only the amino acid residues). It is estimated that the rate of protein synthesis is about 15 amino acid residues/second. Calculate the time needed to complete the synthesis of this protein.

10. a. Which operon terms have the abbreviations I, P, O, Z, Y and A.
    b. What is the order of these regions in the operon DNA?
    c. Indicate what effect a defective mutation (e.g., a premature stop codon or one that would alter the binding of a regulator protein to the DNA sequence) in each of the following regions would have on the expression of the structural genes.
       i. I region
       ii. O region
       iii. P region

11. Name the two possible fates of a cell upon infection with the SV40 virus.

12. Indicate whether the following statements are true or false. If a statement is false, correct it.
    a. The minimum number of adjacent nucleotides in mRNA that can define the genetic code is three.
    b. In *E. coli*, the tRNA^{fmet} combines with N-formylmethionine to form the initial amino acid in the synthesis of a protein.
    c. The initiation of translation in *E. coli* requires the formation of a complete ribosome.
    d. Translocation is the movement of the peptidyl-tRNA from the P site to the A binding site.
    e. Puromycin blocks translation by inhibiting the enzyme peptidyl transferase.
    f. The only amino acids that have unique codons are methionine and tryptophan.
    g. The binding of CAP-cAMP to the promoter site in the *lac* operon inhibits the expression of the *lac* structural genes.

Chapter 20 Protein Synthesis

    h.   Simian virus 40 is a retrovirus with an oncogene in its genome, which encodes the large-T-protein.

    i.   The most common mutation in human cancers is in the gene which encodes for the p21 protein.

ANSWERS TO EXERCISES

1.   a.   amino acid + ATP + tRNA ---> aminoacyl-tRNA + PP$_i$
    b.   Aminoacyl-tRNA synthetase catalyzes this reaction.
    c.   There are two high energy bonds hydrolyzed in this process. ATP is hydrolyzed to yield AMP + PP$_i$. The PP$_i$ is then hydrolyzed to two phosphates.

2.   a.   The amino acid binds to the 3' end of the tRNA.
    b.   The reaction occurs at the 3'-hydroxyl group of the ribose ring of adenosine. Recall that one of the common features in all tRNA molecules is that the last three nucleotides at the 3' end are CpCpA. The amino acid-tRNA linkage formed is an ester bond (refer to Figure 20.5 in the text).

3.   a.   In *E. coli*, the ribosomal subunits are 30S and 50S, with the complete ribosome having an S value of 70.
    b.   Two types of hydrogen bonding interactions determine the start point for protein synthesis. The mRNA interacts with the Shine-Dalgarno sequence in the 3' end of the 16S rRNA and the anticodon sequence of the initiator tRNA^{fmet} interacts with the start codon on mRNA.
    c.   The 5' end of the mRNA interacts with the 3' end of the 16S rRNA.
    d.   The Shine-Dalgarno sequence is the purine-rich sequence, 5'-GGAGGU-3'. It resides on the 5' end of the mRNA, about 10 nucleotides upstream from the initiator codon.
    e.   The corresponding pyrimidine-rich DNA sequence is 5'-ACCTCC-3'.

4.   a.   DNA is synthesized in the 5' ---> 3' direction.
    b.   The DNA strand is read by DNA polymerase in the 3' ---> 5' direction.
    c.   RNA is synthesized in the 5' ---> 3' direction.
    d.   The DNA strand that is read by RNA polymerase is in the 3' ---> 5' direction.

Chapter 20 Protein Synthesis

    e. Protein synthesis begins at the 5' end of the mRNA molecule.

    f. The codon and anticodon triplets are oriented in an antiparallel alignment, with complementary base pairs between the mRNA and the tRNA molecules.

    g. The amino end of the protein is synthesized first.

5. There are no organelles in prokaryotic organisms. Both transcription and translation occur in the cytoplasm. Therefore, after only a small segment of the mRNA has been synthesized, translation can begin immediately.

   In eukaryotic cells, transcription occurs in the nucleus, while translation occurs on the ribosomes in the cytoplasm. The immature mRNA undergoes extensive processing before it leaves the nucleus and is transported to the cytoplasm for protein synthesis. Only after the mRNA interacts with the ribosomes in the cytoplasm can translation begin. Therefore, transcription and translation are separated both spatially and temporally in eukaryotic cells.

6.   a. Translation is the process of synthesizing proteins from the instructions in the genetic material. The molecular messenger that brings the instructions from DNA to the cytoplasm is mRNA. This is the point at which the "nucleotide alphabet" is converted to the "amino acid alphabet" and the molecular communications problem, dealing with two different languages, must be resolved. The nucleotide alphabet has four letters, while the amino acid alphabet has twenty letters. The genetic code is the direct correspondence between these two languages. A triplet of adjacent nucleotides in mRNA defines an amino acid to be incorporated into a protein.

     b.   i. Protein synthesis with poly(UUU) serving as the mRNA will produce a homopolypeptide with only phenylalanine residues in it (i.e., poly(phenylalanine)).

         ii. Translation using poly(GCA) as the mRNA will also produce a homopolypeptide, polyalanine. This result is consistent with the genetic code being made up of triplets of nucleotides.

Chapter 20 Protein Synthesis

      iii.   Translation with poly(AU) will produce a heteropolypeptide, since the genetic code involves a triplet of nucleotides. Therefore, the message in poly(AU) will be a repeating unit of AUA and UAU triplets - AUAUAUAUAUAUAUAUAU...
This will produce poly(ile-tyr).

7.   a.   The codon is read in a 5' --->3' direction in mRNA. The codon-anticodon interact by antiparallel strands, with the anticodon in a 3' ---> 5' direction. The two triplets interact through Watson-Crick base pairing.

      b.   The "wobble" hypothesis refers to the observation that the base at the 5' end of the anticodon has less restrictions on the base pairs it may form. The other two bases in the anticodon are restricted to engaging in only Watson-Crick base pairing with the bases in the codon triplet.

8.   DNA   5'-**GGA**CCA**GTC**ACA**CAT**-3'

       mRNA   5'-**AUG**UGU**GAC**UGG**UCC**-3'
Alternating triplets are shown in bold type for easier reading.

       protein   +H$_3$N-met-cys-asp-trp-ser---

9.   Assume the average molecular weight of an amino acid residue is approximately 105 Daltons. The number of amino acid residues in this protein would be 155,400/105 = 1480. At a rate of translation of 15 residues/second, it would require only about 99 seconds or slightly longer than a minute and one-half to synthesize this enormous protein.

10.   a.   The abbreviations for the following operon terms are,
           I- the structural gene for (production of) the repressor protein
           P- the promoter region
           O- the operator region
           Z, Y and A- the structural genes.

The I, P and O regions are upstream from the structural genes and control the expression of these genes into RNA and the protein products.

Chapter 20 Protein Synthesis

b. The order of the operon elements, from the 5' end to the 3' end of the DNA, is I, P, O, followed by the structural genes, Z, Y and A.

c.
i. A mutation that produces a premature stop codon results in a nonfunctioning protein product. This type of mutation in I would result in the incomplete synthesis of the repressor molecule. If there were no functional repressor available to bind to the operator, synthesis would be continuously turned on.

ii. The operator sequence does not code for a protein. A mutation in the operator may alter the binding of the repressor protein to the operator sequence. If the binding interaction were weaker as a result of the mutation, RNA polymerase might be able to "compete the repressor off" the operator sequence and carry out transcription. This would result in an increase of transcription to some degree.

iii. The promoter sequence does not code for a protein. A mutation in the promoter sequence will alter the extent that RNA polymerase binds to it. If the mutation weakens this interaction, this may result in a decrease in the level of transcription.

11. The virus can infect the cell, produce many copies of the virus, and as a result, the burst the cell (i.e., lyse and die). The viruses will be free to infect more cells and to cause more cell deaths. Alternatively, the virus can integrate its DNA into the genome of the host cell. In this case, the infected cell survives, but loses control of its own growth. The cell becomes a cancer cell. A cancer cell is one which exhibits unregulated growth.

12.
a. True.
b. False. The t-RNA^{fmet} combines with methionine, which is subsequently modified to produce N-formylmethionine.
c. True.
d. False. The movement is from the A binding site to the P site.
e. False. Puromycin blocks translation by either (1) binding to the P site and blocking translocation or (2) being

Chapter 20 Protein Synthesis

incorporated into the peptide at the carboxyl end of the growing peptide chain. The peptidyl-puromycin then dissociates from the ribosome as a prematurely terminated protein.

f. True.
g. False. The CAP-cAMP complex must bind to the promoter site for transcription to occur.
h. False. Simian virus 40 is a DNA virus with an oncogene in its genome, which encodes the large-T-protein.
i. The most common mutation in human cancers is in the gene which encodes the p53 protein.